Computational Spectroscopy of Polyatomic Molecules

This book provides a detailed description of the modern variational methods available for solving the nuclear motion Schrödinger equation to enable accurate theoretical spectroscopy of polyatomic molecules. These methods are currently used to provide important molecular data for spectroscopic studies of atmospheres of astronomical objects including solar and extrasolar planets as well as cool stars. This book has collected descriptions of quantum mechanical methods into one cohesive text, making the information more accessible to the scientific community, especially for young researchers, who would like to devote their scientific career to the field of computational molecular physics.

The book addresses key aspects of the high-accuracy computational spectroscopy of medium-size polyatomic molecules. It aims to describe numerical algorithms for the construction and solution of the nuclear motion Schrödinger equation with the central idea of the modern computational spectroscopy of polyatomic molecules to include the construction of the complex kinetic energy operators (KEOs) into the computation process of the numerical pipeline by evaluating the corresponding coefficients of KEO derivatives on-the-fly. The book details the key aspects of variational solutions of the nuclear motion Schrödinger equations targeting high accuracy, including the construction of rotational and vibrational basis functions, coordinate choice, molecular symmetry as well as aspects of intensity calculations and refinement of potential energy functions. The goal of this book is to show how to build an accurate spectroscopic computational protocol in a pure numerical manner of a general black-box-type algorithm.

This book will be a valuable resource for researchers, both experts and not experts, working in the area of the computational and experimental spectroscopy; PhD students and early-career spectroscopists who would like to learn the basics of modern variational methods in the field of computational spectroscopy. It will also appeal to astrophysicists and atmospheric physicists who would like to assess data and perform calculations themselves.

Computational Spectroscopy of Polyatomic Molecules

Sergey Yurchenko

CRC Press is an imprint of the
Taylor & Francis Group, an **informa** business

First edition published 2023
by CRC Press
6000 Broken Sound Parkway NW, Suite 300, Boca Raton, FL 33487-2742

and by CRC Press
4 Park Square, Milton Park, Abingdon, Oxon, OX14 4RN

CRC Press is an imprint of Taylor & Francis Group, LLC

ISBN: 978-1-498-76119-2 (hbk)
ISBN: 978-1-032-43372-1 (pbk)
ISBN: 978-0-429-15434-8 (ebk)

DOI: 10.1201/9780429154348

Typeset in CMR10 font
by KnowledgeWorks Global Ltd.

Publisher's note: This book has been prepared from camera-ready copy provided by the authors.

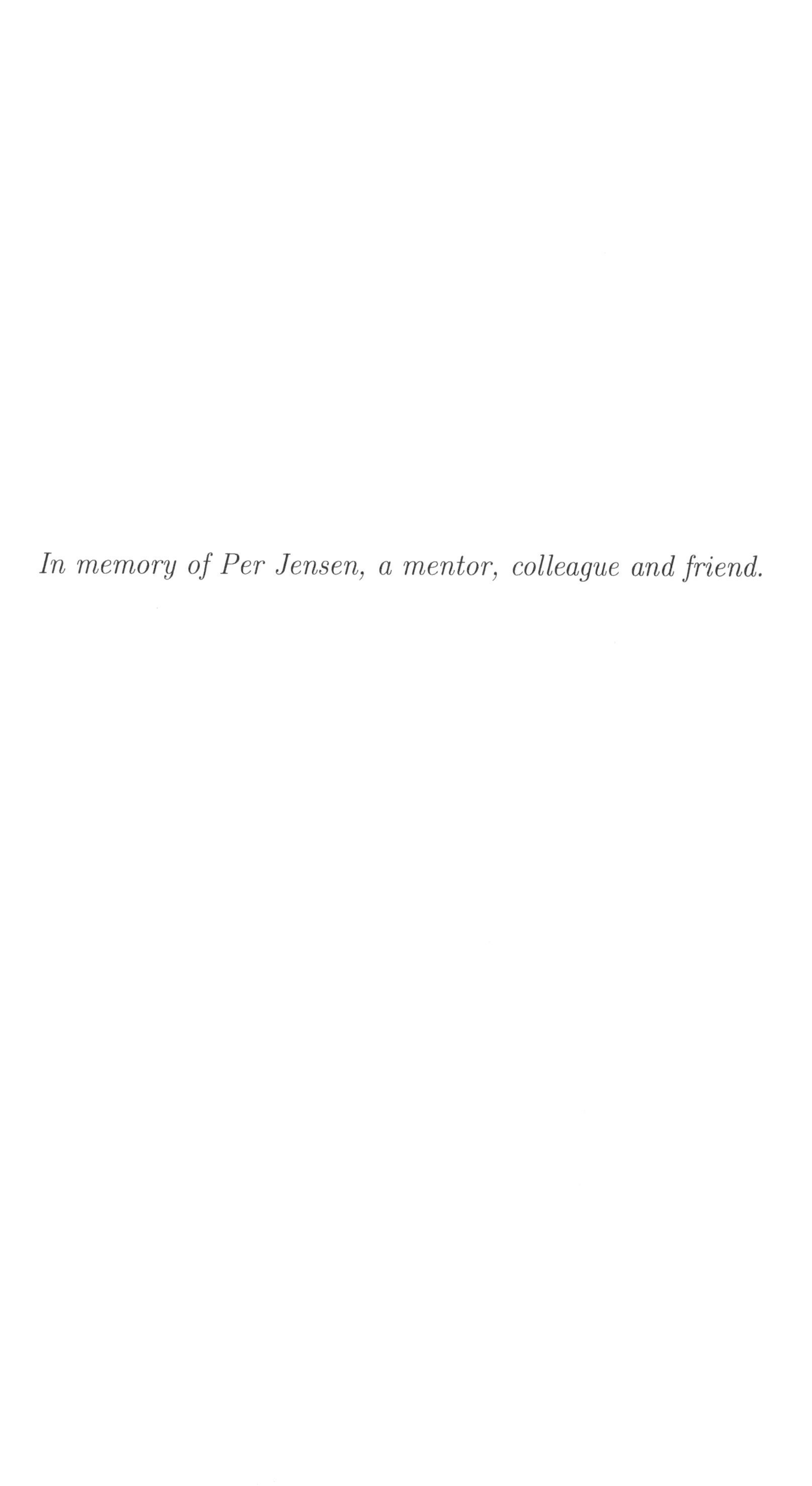

In memory of Per Jensen, a mentor, colleague and friend.

Contents

Preface

The book aims to describe numerical algorithms for the construction and solution of the nuclear motion Schrödinger equations of (small or medium) polyatomic molecules mainly designed for high-accuracy spectral applications. High standards of the modern experimental spectroscopy have pushed theory to its limit demanding more and more extensive (and expensive) computations. The key idea that revolutionised the modern computational spectroscopy of polyatomic molecules in the past two decades was to include the construction of the complex kinetic energy operators (KEOs) into the computation process of the numerical pipeline by numerically evaluating the corresponding coefficients of KEO derivatives. The goal of this book is to show how to build a spectroscopic computational protocol in a pure numerical manner of a general black-box-type algorithm.

This book is not a full coverage of the research field dedicated to the accurate variational solution of rotation-vibrational Schrödinger equations for polyatomic molecules. This would be impossible considering the diversity of methods explored by the community. The subject of the book is on the usage of the so-called finite basis representation and sum-of-products representations of the molecular Hamiltonian operator, with the special emphasis on black-box algorithms and their efficiency for high-performance computations. To some extent, materials described in the book have been implemented, tested and used in the program TROVE (Yurchenko et al., 2007), developed by the author in cooperation with Per Jensen, Walter Thiel and Andrey Yachmenev and thus reflect the personal research path of the author via the methods he developed and the programs he wrote. These developments have been used in dozens of accurate spectroscopic applications. Thus, this book is for those who are interested in high-resolution rotational-vibrational spectroscopy of polyatomic molecules, especially those who are involved in method development, implementation and high-performance computing of spectra of polyatomics. Different choices of the internal coordinates and molecular frames are well illustrated. These concepts play central roles in the solution of nuclear motion Schrödinger equations.

I owe immensely to my many teachers, colleagues and friends who influenced and shaped my scientific path and legacy, listed here in the order of appearance in my life: Yurii S. Makushkin, Aleksander V. Shapovalov, Oleg N. Ulenikov, Per Jensen, Miguel Carvajal Zaera, Philip R. Bunker, Walter Thiel, Andrey Yachmenev and Jonathan Tennyson.

The inspiration for my research and this book in particular was drawn from the work of many great theoretical spectroscopists, most notably of Tucker Carrington, Attila Czászár, Lauri Halonen, Jon T. Hougen, David Lauvergnat, Renato Lemus,

Edit Mátyus, André Nauts, Andrei Nikitin, Anatoly I. Pavlyuchko, Georg Ole Sørensen, Brian T. Sutcliffe and James K.G. Watson.

I have benefited from the comments and suggestions of many colleagues, who read drafts of the book. I would like to thank in particular Jonathan Tennyson, Alec Owens, Thomas Mellor, Anthony E. Lynas-Gray, Barry Mant, Andrei Sokolov, Wilfrid Somogyi, Samuel Wright and Ryan Brady.

Finally, I thank my wife Olga and daughter Asya. Without their help, support, and encouragement, this book would never have been completed.

CHAPTER 1

Introduction

The modern variational calculations of molecular spectra by solving the nuclear motion Schrödinger equation are extremely computationally involved. Based on sophisticated computer programs, they require ever-growing computer power. To model experimental data with adequate accuracy, second-order differential equations for the motion of nuclei have to be solved in the rotating frame, with the differential part (kinetic energy term) becoming very complex even for small polyatomic (e.g. tetratomic) molecules. The common approach is to derive the kinetic energy operator (KEO) analytically using a computer algebra program and then implement it for large-scale numerical calculations. This involves a transformation of the second-order differential form from the laboratory-fixed frame to the rotating, molecular frame. This approach, however, proved to be difficult for larger molecules (with more than four atoms) even with modern analytic software and computer resources involved. It has been recently realised that it is possible and even more efficient to include the construction of the KEO into the computational pipeline and thus to avoid any analytic work. This realisation and subsequent code developments led to a breakthrough in computational spectroscopy of polyatomic molecules. This book is a reflection of the author on modern computational techniques based on the black-box, numerical paradigm.

In this book we will deal with quantum-mechanical computational methods applied to the vibrational and rotational motion of medium-size polyatomic molecules (three to ten atoms). We will assume a molecule to be in an isolated, singlet (spin-free) electronic state (e.g. ground) such that any interaction with other electronic states can be ignored and only the rotation-vibration problem can be considered. This is a very common situation for many polyatomic molecules including those important for atmospheric applications. We will also neglect the hyperfine structure or at least the interaction of the nuclear spin with the ro-vibrational degrees of freedom, which is usually a good approximation for all but a few special cases (e.g. ortho-para conversions).

We will assume that the electronic problem has been fully addressed under the Born-Oppenheimer approximation and resulted in a potential energy surface (PES) and a electronically averaged dipole moment surface (DMS). A PES represents the internal energy of the molecule at different nuclear geometries, i.e. different

DOI: 10.1201/9780429154348-1

arrangements of the nuclei. We also assume that such a PES is readily available for energy calculations as a potential energy function (PEF) of internal, vibrational coordinates. The rotation-vibration energies (and corresponding wavefunctions) of a molecule are obtained as a solution of the Schrödinger equation for the given PES. The molecular DMS is then combined with the wavefunctions to calculate transition intensities and to simulate ro-vibrational spectra. Under these assumptions, the internal (electronic, rotation, vibration, nuclear-spin) wavefunction of the molecule is given by

$$\Psi(\boldsymbol{r}_{\rm e}, \boldsymbol{R}_n, \boldsymbol{\sigma}_{\rm ns}) = \Psi_{\rm e}^{(\boldsymbol{R}_n)}(\boldsymbol{r}_{\rm e}) \times \Psi_{\rm rv}(\boldsymbol{R}_n) \times \Psi_{\rm ns}(\boldsymbol{\sigma}_{\rm ns}),$$

where $\boldsymbol{r}_{\rm e}$, $\boldsymbol{R}_n$ and $\boldsymbol{\sigma}_{\rm ns}$ are electronic, nuclear and nuclear-spin coordinates, respectively; $\Psi_{\rm e}^{(\boldsymbol{R}_n)}(\boldsymbol{r}_{\rm e})$ is an electronic wavefunction, parameterised with the nuclear coordinates $\boldsymbol{R}_n$; $\Psi_{\rm rv}(\boldsymbol{R}_n)$ and $\Psi_{\rm ns}(\boldsymbol{\sigma}_{\rm ns})$ are the ro-vibrational and nuclear-spin wavefunctions, respectively. In this book we will only concern ourselves with the ro-vibrational nuclear wavefunction $\Psi_{\rm rv}(\boldsymbol{R}_n)$, associated energies $E_{\rm rv}$, nuclear coordinates $\boldsymbol{R}_n$ and, most importantly, with the ro-vibrational Hamiltonian operator $\hat{H}^{\rm rv}$ that 'glues' them together:

$$\hat{H}^{\rm rv}\Psi^{\rm rv} = E^{\rm rv}\Psi^{\rm rv}, \tag{1.1}$$

which is a *time-independent* Schrödinger equation for the ro-vibrational motion of nuclei.

The nuclear motion is a combination of translation, rotation and vibrations of the nuclei. For an N-atomic molecule, there are three translational, three rotational and $3N-6$ vibrational degrees of freedom (for a non-linear molecule). The three translational degrees of freedom can always be excluded for a free molecule in isomorphic space by placing the coordinate system at the centre-of-mass. The rotational degrees of freedom cannot be fully decoupled from the molecular vibrational motion because of the centrifugal distortion effects, and we thus need to consider a $3N-3$ dimensional rotation and vibration of the nuclei quantum-mechanically. The ro-vibration Hamiltonian $\hat{H}^{\rm rv}$ is a sum of the KEO $\hat{T}$ and the PEF V:

$$\hat{H}^{\rm rv} = \hat{T} + V.$$

We will be not interested in the continuum solution, i.e. the eigenfunctions will be constructed assuming the boundary conditions for fully bound, localised states:

$$\lim_{|\boldsymbol{r}_{nm}|\to\infty} |\Psi_i^{\rm rv}|^2 \to 0, \tag{1.2}$$

where $|\boldsymbol{r}_{nm}|$ represents the distance between any nuclei n and m in the molecule. This boundary condition will lead to the discrete solutions $\Psi_i^{\rm rv}$ and $E_i^{\rm rv}$ (i is an index to number them), i.e. they can be satisfied only for specific energy values $E_i^{\rm rv}$. Any pre-dissociation effects will be also ignored, i.e. we will assume that the

PES of the molecule in question is deep enough to accommodate a sufficient number of bound states, for which the interaction with the continuum or other electronic states (non-adiabatic effects) can be disregarded. This condition also means that the following integral is finite:

$$\int_r |\Psi_i^{\mathrm{rv}}|^2 dr,$$

where r represents the coordinate configuration space.

Except for a few simple cases, Schrödinger equations (1.1) describing motion of nuclei do not have analytically exact solutions and therefore have to be solved either using perturbation theory or numerically. Our method of choice is the numerical solution of the time-independent Schrödinger equation using the variational approach. In fact, systems containing more than 6–10 nuclei are challenging for any numerical methods, including variational approaches, high-order perturbation theory methods (e.g. contact transformations), time-dependent propagation methods or different combinations of these three. Therefore, efficient approximations and models are crucial for solutions of these systems.

In the variational approach the unknown eigen-solution Ψ is represented as a linear combination of some known, so-called basis functions ϕ_k from a complete set:

$$\Psi = \sum_{k=1}^{\infty} C_k \phi_k \tag{1.3}$$

with unknown coefficient C_k. The basis functions are chosen to satisfy the bound-states boundary condition in Eq. (1.2). The coefficients are 'varied' to obtain the solution which minimises the energy

$$E = \frac{\int_r \Psi^* \hat{H} \Psi dr}{\int_r \Psi^* \Psi dr}.$$

Without loss of generality, at least as far as the subject of this book is concerned, we also assume that the basis set functions ϕ_k are ortho-normal:

$$\int_r \phi_j^* \phi_k dr = \delta_{jk},$$

where δ_{jk} is the Kronecker delta and * indicates complex conjugation. In practice, the basis set expansion in Eq. (1.3) is truncated at some $k_{\max}$ and the choice of the basis functions ϕ_k becomes an important factor affecting the size of the problem. Assuming that the matrix elements

$$H_{jk} \equiv \int_r \phi_j^* \hat{H} \phi_k dr \tag{1.4}$$

can be evaluated, we arrive at the numerical formulation of the Schrödinger equation in a form of an eigen-problem given by

$$\boldsymbol{HC} = E\boldsymbol{C}, \tag{1.5}$$

where $\boldsymbol{H}$ and $\boldsymbol{C}$ are the Hamiltonian matrix and a (column) vector of eigen-coefficients, respectively. The solution of Eq. (1.5) in the form of a truncated (convergent) expansion in Eq. (1.3) is equivalent to finding a finite $k_{\rm max} \times k_{\rm max}$ unitary transformation $\boldsymbol{U}$ to a diagonal representation of $\boldsymbol{H}$, where the diagonal elements are eigenvalues E_i. In this book we will rely on the existence of efficient numerical methods for diagonalisation of large matrices and thus will not discuss diagonalisations in detail. Instead we will focus on the construction of the KEO in terms of the coordinates, suitable for accurate, large-scale variational solutions as well as on how to facilitate this solution by choosing and building appropriate basis sets.

One of the main topics of this book is the derivation of a KEO, i.e. a transformation from the simple Cartesian form

$$\hat{T} = -\frac{\hbar^2}{2}\sum_i \frac{1}{m_i}\nabla_i^2,$$

where m_i are the corresponding nuclear (or atomic) masses, i is the atomic number and $\nabla_i^2 = \frac{\partial^2}{\partial r_i^2}$ is a Laplace operator for the nuclear coordinate $r_i = \{x_i, y_i, z_i\}$, to the form represented by the vibrational coordinates in a rotated frame, with the focus on pure numerical formulations of this transformation as well as of the solution of Eq. (1.1).

To summarise, if a typical ro-vibrational project of the 20th century was based on the derivation of a ro-vibrational KEO analytically, possibly using computer algebra and implementation as a numerical (Fortran or C) program, customised to work only for specific molecular systems, the focus of the 21st century, which is sometimes referred to as the fourth age of the computational chemistry (Császár et al., 2012), is the development of general, black-box-type algorithms, which include the construction of KEOs into the numerical calculational procedure. The book presents implementation solutions for building numerical algorithms for efficient nuclear motion, ro-vibrational calculations.

CHAPTER 2

Coordinates choice

In this chapter we discuss the importance of the coordinates choice for an accurate ro-vibrational description of the molecular motion. The concept of the moving molecular frame is introduced. Some popular coordinate systems are considered, including normal modes, valence, Jacobi, Radau and linearised coordinates as well as Eckart, Jacobi and principle axis system frames. Examples of specific molecular systems (rigid and non-rigid) are given.

2.1 TRANSFORMATION FROM THE LABORATORY TO MOLECULAR FRAMES

We start from the observation that the description of ro-vibrational motion of a polyatomic molecule benefits a lot from the coordinate system attached to the translating and rotating molecule, with the vibrational motion seen relative to this moving, body-fixed (BF) system (frame); see Fig. 2.1. The translational motion in the absence of an external field is fully independent from the other degrees of freedom and can therefore be removed from consideration. The rotational and vibrational degrees of freedom cannot be fully separated, but it is possible to at least reduce their coupling by selecting the corresponding rotating frame. In fact this is also important for practical applications. Smaller inter-mode couplings improve the quality of the basis set in variational calculations and thus reduce the size of the calculations. Physically intuitive basis sets are especially important for the description of high excitations, both rotational and vibrational.

Let us now define a laboratory-fixed (space fixed) reference frame with the right-handed axes XYZ and place the observer in its centre; see Fig. 2.1. The classical kinetic energy operator (KEO) of a molecule in the laboratory frame (LF) XYZ and Cartesian coordinates ($\boldsymbol{R}_n = \{R_{n\alpha}\}$, $n = 1, \ldots, N$) is given by

$$T = \frac{1}{2}\sum_{n=1}^{N}\frac{1}{m_i}p_n^2. \tag{2.1}$$

DOI: 10.1201/9780429154348-2

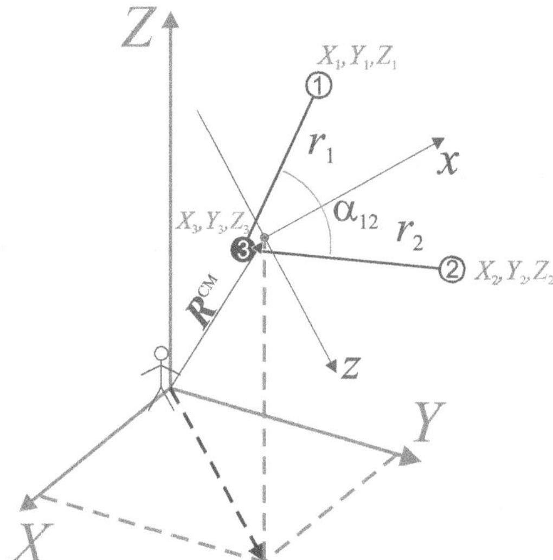

Figure 2.1 Coordinate transformation from the laboratory-fixed frame XYZ ($\boldsymbol{R}_1$, $\boldsymbol{R}_2\boldsymbol{R}_3$) to the BF frame xyz for an XY_2 molecule for an arbitrary instantaneous configuration as an example. The coordinates representing the body-fixed frame are (i) the centre-of-mass position vector $\boldsymbol{R}^{\mathrm{CM}}$ describing the translational motion of the nuclear masses, (ii) the Euler coordinates ϕ, θ and χ (not shown) describing the orientation of xyz relative to XYZ and the three valence coordinates r_1, r_2 and α_{12} describing the vibrational degrees of freedom. Depicted is a bond frame, with the x axis parallel to the bisector of α_{12} and the y axis orthogonal to the plane containing XY_2 (indicated by a circle with a point at the centre).

The quantum-mechanical form, also in a Cartesian representation, is then given by

$$\hat{T} = \frac{1}{2}\sum_{n=1}^{N}\frac{1}{m_n}\hat{p}_n^2 = -\frac{\hbar^2}{2}\sum_{n=1}^{N}\frac{1}{m_n}\nabla_n^2. \tag{2.2}$$

Here, nucleus n ($= 1, 2, 3, \ldots, N$) has mass m_n; R_{nX}, R_{nY} and R_{nZ} are the Cartesian coordinates in the LF axis system XYZ; and $\nabla_n^2 = \partial^2/\partial R_{nX}^2 + \partial^2/\partial R_{nY}^2 + \partial^2/\partial R_{nZ}^2$.

Expressed in terms of the Cartesian coordinates of the nuclei in a LF axis system, the Schrödinger equation for the translation, rotation and vibration of a polyatomic molecule has a simple form:

$$\left(-\frac{\hbar^2}{2}\sum_{i=1}^{N}\frac{1}{m_n}\nabla_n^2 + V\right)\Psi_{\mathrm{trv}} = E_{\mathrm{trv}}\Psi_{\mathrm{trv}}. \tag{2.3}$$

In this equation, V is the Born-Oppenheimer potential energy function of the nuclei and Ψ_{trv} is the translation-rotation-vibration wavefunction with the associated translation-rotation-vibration energy E_{trv}. The potential energy function does not depend on the translational or rotational motions and is a function of the relative displacement of the nuclei.

Even though the Schrödinger equation in Eq. (2.3) has a mathematically simple form, it is badly suited for actual solution. A more suitable coordinate choice is associated with molecule-fixed axis systems xyz (referred to also as BF axis frames) with origin at the nuclear centre-of-mass of the molecule (CM) with the translational motion described by a position vector $\boldsymbol{R}^{\rm CM}$ of CM relative to the lab frame XYZ. An example of the xyz frame in the case of the XY_2 type molecule is also shown in Fig. 2.1. The rotational motion of the molecule as a whole is represented by the orientation of the xyz frame relative to the LF frame system XYZ, which in a general case of a non-linear molecule requires a set of three angles (two for a linear system). The molecular vibrations are then represented by the relative positions of N nuclei in a non-inertial or BF frame xyz using some suitable $3N-6$ coordinates (or $3N-5$ for a linear molecule), where $3N$ is the total number of degrees of freedom and 6 (5) is the number of translational plus rotational degrees of freedom. In the following, unless specified, the expression $3N-6$ will be used to refer to the generic quantity describing the total number of the vibrational degrees of freedom, which in the case of a linear molecule is assumed to be $3N-5$.

The coordinate transformation from the LF system represented by $3N$ Cartesian coordinates $R_{n,A}$ $(A = X, Y, Z)$ to a BF frame can be formulated as

$$R_{n,A} = R_A^{\rm CM} + \sum_{\alpha=x,y,z} S_{A,\alpha}(\theta,\phi,\chi)\, r_{n,\alpha}^{\rm BF}(\boldsymbol{\xi}), \tag{2.4}$$

where $\boldsymbol{R}^{\rm CM} = \{X^{\rm CM}, Y^{\rm CM}, Z^{\rm CM}\}$ is a three-dimensional (3D) position vector of the CM; $\boldsymbol{S} = \{S_{A,\alpha}(\theta,\phi,\chi)\}$ is the 3×3 unitary matrix of the directional cosines describing the rotation of the BF system xyz relative to the LF system XYZ as function of the three angles θ,ϕ and χ (typically Euler angles, see Section 2.2.2 for definition); and $\boldsymbol{r}_n^{\rm BF}(\boldsymbol{\xi})$ are the $3N$ Cartesian coordinates describing the positions of the nuclei in the moving xyz coordinate system, which effectively depend on $M = 3N-6$ independent internal (vibration) degrees of freedom. The coordinate transformation given in Eq. (2.4) has a dramatic effect on the form of the KEO, as will be described in detail in Chapter 3, with $\boldsymbol{\xi}$-depended 'inverse masses' and a pseudo-potential function or Watson term as part of the KEO.

The choice of the M vibrational coordinates $\{\xi_1, \xi_2, \ldots, \xi_M\}$ is system dependent. The most common choices include the normal-mode coordinates (or other rectilinear coordinates such as linearised) and geometrically defined coordinates (e.g. valence coordinates such as bond lengths, inter-bond and dihedral angles). Thus the transformation from $\boldsymbol{R}_n$ to $\{X^{\rm CM}, Y^{\rm CM}, Z^{\rm CM}\}, \{\theta,\phi,\chi\}, \{\xi_1,\xi_2,\ldots,\xi_M\}$ in Eq. (2.4) of the $3N$ BF Cartesian coordinates requires the following $3N$ constraints, also discussed in detail in the next section: three centre-of-mass conditions for the translation coordinates $\boldsymbol{R}$; three rotational conditions for the definition of the BF axes xyz in an instantaneous configuration of the N nuclei; $3N-6$ conditions for the definition of the vibrational coordinates ξ_i $(i = 1, 2, \ldots, M)$. In the example shown in Fig. 2.1, the three internal, vibrational coordinates are the bond lengths r_1 and r_2 and the inter-bond angle α.

2.2 DEFINING THE COORDINATE TRANSFORMATION

2.2.1 Three translational conditions

The position of the nuclear centre-of-mass is given by the CM vector

$$\boldsymbol{R}^{\rm CM} = \frac{\sum_{n=1}^{N} \boldsymbol{R}_n m_n}{\sum_{n=1}^{N} m_n},$$

where $\sum_{n=1}^{N} m_n \equiv M$ is the total mass of the nuclei. Let us first transform the LF frame to the body frame $xyz^{\rm BF}$, which is centred at $\boldsymbol{R}^{\rm CM}$ and parallel to XYZ (see Fig. 2.1)

$$\boldsymbol{R}_n = \boldsymbol{R}^{\rm CM} + \boldsymbol{r}_n^{\rm BF},$$

with N position vectors $\boldsymbol{r}_n^{\rm (BF)}$ satisfying the three centre-of-mass conditions for any instantaneous positions of the nuclei and orientation of the molecule:

$$\sum_{n=1}^{N} m_n \boldsymbol{r}_n^{\rm BF} = 0 \tag{2.5}$$

or for individual modes:

$$\sum_{n=1}^{N} m_n r_{n,x}^{\rm BF} = 0, \tag{2.6}$$

$$\sum_{n=1}^{N} m_n r_{n,y}^{\rm BF} = 0, \tag{2.7}$$

$$\sum_{n=1}^{N} m_n r_{n,z}^{\rm BF} = 0. \tag{2.8}$$

It can be easily shown that using this transformation in the Nabla operator ∇ in Eq. (2.3), the 3D translation motion is separable from the rest degrees of freedom in the KEO:

$$\hat{T} = \hat{T}^{\rm CM} + \hat{T}^{N},$$

where

$$\hat{T}^{\rm CM} = -\frac{\hbar^2}{2M} \nabla^2_{\rm CM}$$

and

$$\hat{T}^{\rm N} = -\frac{\hbar^2}{2} \sum_{i=2}^{M} \frac{\nabla_i^2}{m_i} + \frac{\hbar^2}{2M} \sum_{i,i'=2}^{M} \nabla_i \nabla_{i'}. \tag{2.9}$$

Here, the coordinates and conjugate momenta of the 1st particle were eliminated and replaced by the centre-of-mass coordinates and momenta, respectively. In the following, however, the separation of the CM coordinates will be done as a part of a generalised transformation starting from Eq. (2.2) and not via Eq. (2.9).

Since the potential energy function does not depend on the centre-of-mass coordinates $\boldsymbol{R}^{\mathrm{CM}}$ and the CM kinetic energy part is fully uncoupled from the other $3N-3$ coordinates $\boldsymbol{r}^{\mathrm{BF}}$, they are also separable in the nuclear eigenfunctions:

$$\Psi(\boldsymbol{R}^{\mathrm{CM}}, \boldsymbol{r}_n^{\mathrm{BF}}) = \Psi(\boldsymbol{R}^{\mathrm{CM}})\Psi(\boldsymbol{r}_n^{\mathrm{BF}})$$

and so are the corresponding eigenvalues

$$E = E^{\mathrm{CM}} + E^{\mathrm{RV}}.$$

The CM part $\Psi(\boldsymbol{R}^{\mathrm{CM}})$ of the eigenfunction of a field-free Hamiltonian in isomorphic space has a trivial solution of a plane wave

$$\Psi(\boldsymbol{R}^{\mathrm{CM}}) = Ce^{i(\boldsymbol{k}\cdot\boldsymbol{R}^{\mathrm{CM}})}$$

and is therefore usually ignored. Here $\boldsymbol{k} = \boldsymbol{P}/\hbar$ is the wave vector associated with the translational energy of the molecule

$$E^{\mathrm{CM}} = \frac{\hbar^2|\boldsymbol{k}|^2}{2M}$$

with the position at $\boldsymbol{R}^{\mathrm{CM}}$.

The solution for the $3N-3$ dimensional problem involves the rotational and vibrational motions and is significantly less trivial.

2.2.2 Three rotational conditions

The next step is to define the orientation of the xyz frame for an arbitrary instantaneous configuration of nuclei. In principle, any molecular frame attached to the CM and defining the orientation of the xyz axes relative to the XYZ frame for a given instantaneous configuration should be applicable. Let us consider an example of a diatomic molecule, shown in Fig. 2.2, with the coordinate centre at the centre-of-mass. It is natural here to place the z axis along the molecular bond, which in this case requires only two angles to fully define the orientation of the molecule relative to the lab-frame: polar θ and azimuthal ϕ, which represent the common spherical polar system. The orientation of the x and y BF axes is, however, ambiguous for a linear molecule since the rotation around the z axis is undefined. Thus for the full description of the two nuclei of a diatomic molecule ($N = 2$) with six Cartesian coordinates in the lab-frame, we select three CM coordinates $X^{\mathrm{CM}}, Y^{\mathrm{CM}}$ and Z^{CM} and two angles θ and ϕ. The 6th degree of freedom is naturally chosen as the bond length, i.e. the distance r between the two nuclei.

Let us now consider a non-linear triatomic molecule of the type XY_2, e.g. H_2O, which is a planar object, forming an isosceles triangle at equilibrium. The intuitive choice of the vibrational degrees of freedom which defines an arbitrary, instantaneous configuration of nuclei is the valence coordinates: the inter-bond angle

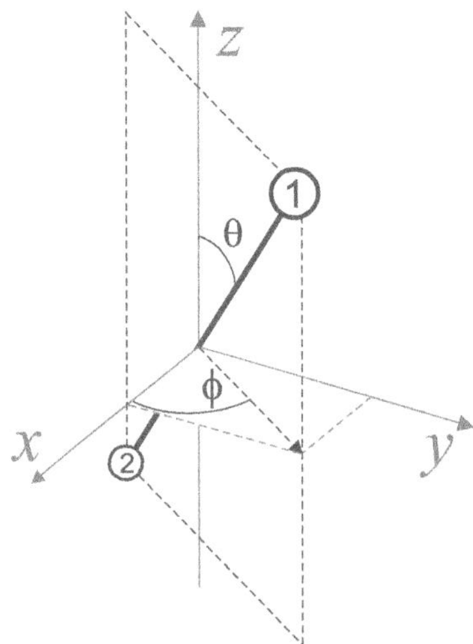

Figure 2.2 Description of the rotation of a diatomic molecule relative to the laboratory-fixed frame.

$\alpha \equiv \alpha_{12}$ between the molecular bonds and their lengths r_1 and r_2 (see Fig. 2.3). We start by centring xyz at CM (three conditions in Eq. (2.6)); see Fig. 2.1. We can also quite naturally define one of the molecular axes (e.g. y) to be orthogonal to the molecular plane. There is, however, no unique choice for the other two axes, at least in the case of an arbitrary position of the nuclei. One of the common choices is to put one of the axes (e.g. x) parallel to the line bisecting the inter-bond angle as in Fig. 2.3. Placing the z axis also in the molecular plane and orthogonal to x (in the right-handed sense) completes the definition of the molecular frame xyz.

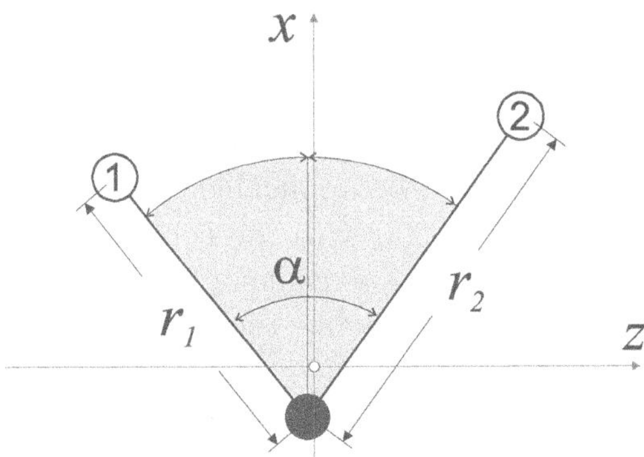

Figure 2.3 Description of a bisector frame for an XY_2 triatomic molecule. The molecular coordinate system is centred at the nuclear centre-of-mass with the x axis parallel to the bisector of the inter-bond angle α.

In order to define the relative orientation of the xyz frame (for non-linear case), we now need to use three angles, θ $(0\ldots\pi)$, ϕ $(0\ldots 2\pi)$ and χ $(0\ldots 2\pi)$. The angles θ and ϕ can be defined as before, i.e. as polar and azimuthal angles specifying the orientation of the z axis relative to XYZ (in the positive direction), see Fig. 2.4, while the angle χ conventionally (Wilson et al., 1955) is chosen to specify the direction of the x axis (or xy plane) measured from the positive half of the node line ON which marks the intersection of the X, Y and x, y. These angles θ, ϕ and χ are called Euler angles, a term that will be used in the rest of the book. The construction can also be understood as follows: in order to reorient the XYZ axes to xyz defined using (ϕ, θ, χ), we first rotate X and Y by an angle ϕ about Z into X' and Y'; we then rotate X' and Z by an angle θ about Y' into X'' and z; finally we rotate X'' and Y' by an angle χ about z into x and y. Note that when the molecule is in a linear configuration this frame turns into the frame of a diatomic with the molecule lying along the z-axis and two angles θ and ϕ as in Fig. 2.2.

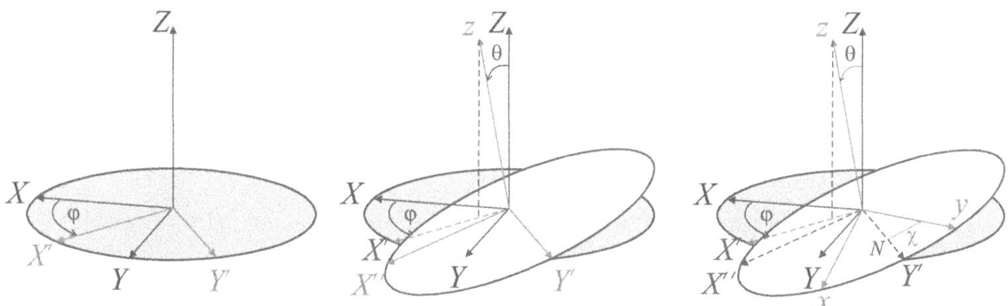

Figure 2.4 The Euler angles describing the orientation of the molecular (x, y, z) axis from the laboratory-fixed (X, Y, Z) axis. Here, θ is the angle between the Z and z axes, ϕ is the angle from X to the projection of z on the $X - Y$ plane, and χ is the angle between the node line ON and the y-axis, where O is the origin of both axes and ON defines the node line, which is the intersection of the $X - Y$ and the $x - y$ planes. ON is also perpendicular to both the z and Z axes. χ is therefore the azimuthal angle about the z-axis.

The actual transformation of the Cartesian coordinate vector from an XYZ coordinate system to a xyz representation for any atom i is given by the directional cosine 3×3 matrix:

$$\begin{pmatrix} x_i \\ y_i \\ z_i \end{pmatrix} = \begin{pmatrix} \lambda_{xX} & \lambda_{xY} & \lambda_{xZ} \\ \lambda_{yX} & \lambda_{yY} & \lambda_{yZ} \\ \lambda_{zX} & \lambda_{zY} & \lambda_{zZ} \end{pmatrix} \begin{pmatrix} X_i \\ Y_i \\ Z_i \end{pmatrix}, \tag{2.10}$$

where the matrix elements $\lambda_{\alpha A}$ are given by

$$\begin{aligned} \lambda_{xX} &= \cos\theta\cos\phi\cos\chi - \sin\phi\sin\chi, \\ \lambda_{yX} &= -\cos\theta\cos\phi\sin\chi - \sin\phi\cos\chi, \\ \lambda_{zX} &= \sin\theta\cos\phi, \end{aligned} \tag{2.11}$$

$$\begin{aligned} \lambda_{xY} &= \cos\theta\sin\phi\cos\chi + \cos\phi\sin\chi, \\ \lambda_{yY} &= -\cos\theta\sin\phi\sin\chi + \cos\phi\cos\chi, \\ \lambda_{zY} &= \sin\theta\sin\phi, \end{aligned} \tag{2.12}$$

$$\begin{aligned}
\lambda_{xZ} &= -\sin\theta\cos\chi, \\
\lambda_{yZ} &= \sin\theta\sin\chi, \\
\lambda_{zZ} &= \cos\theta.
\end{aligned} \tag{2.13}$$

The corresponding rotational angular momentum operators $\hat{J}_x$, $\hat{J}_y$ and $\hat{J}_z$ are given by

$$\hat{J}_x = \sin\chi\,\hat{P}_\theta - \csc\theta\cos\chi\,\hat{P}_\phi + \cot\theta\cos\chi\,\hat{P}_\chi, \tag{2.14}$$

$$\hat{J}_y = \cos\chi\,\hat{P}_\theta + \csc\theta\sin\chi\,\hat{P}_\phi - \cot\theta\sin\chi\,\hat{P}_\chi, \tag{2.15}$$

$$\hat{J}_z = \hat{P}_\chi, \tag{2.16}$$

expressed in terms of the momentum operators

$$\hat{P}_\alpha = -i\hbar\frac{\partial}{\partial\alpha}, \quad \alpha = \{\chi, \theta, \phi\}. \tag{2.17}$$

The angular momentum operators $\hat{J}_\alpha$ ($\alpha = x, y, z$) satisfy the anomalous commutation properties

$$[\hat{J}_\alpha, \hat{J}_\beta] = -i\hbar\hat{J}_\gamma,$$

where α, β, γ are x, y, z or their cyclic permutations.

2.3 MOLECULAR FRAMES

In this section, examples of molecular frames are presented and discussed.

2.3.1 Bond frame

A body frame xyz system can always be chosen based on some geometrical arguments (see below) analogous to the examples of diatomic and XY_2 molecules from the previous section. Let us consider two main criteria choosing a frame: a more efficient separation of the rotation and vibrational degrees of freedom and simplicity of the KEO. At equilibrium, it is common to use the principal axes system (PAS) for the xyz frame, characterised by a diagonal moment of inertia tensor. For any arbitrary instantaneous configuration of the vibrating nuclei, the PAS is not always the optimal choice for xyz (see Section 2.3.3) due to ensuing singularities. Yet, it is logical to expect that even a deformed, vibrating structure would still approximately follow the equilibrium or PAS frame, at least for relatively small distortions.

Let us start with a simple example of a non-symmetric triatomic XYZ-type molecule HCN. At its equilibrium HCN is a linear molecule and is therefore a symmetric top. It is physically intuitive to expect a stable rotation around the heavier bond (CN). We therefore can choose the z axis to be parallel to the C–N bond, with the x in the molecular plane and y orthogonal to x and z, in the right-handed sense.

For molecules with heavy C–C backbones, e.g. C_2H_2, C_2H_4 and C_2H_6, placing the z axis parallel to the C–C bond is also a physically intuitive choice (but always with the origin at the centre-of-mass). Similar examples with the z axis parallel to some backbone-like bonds include H_2CO (parallel to C–O), H_2O_2 (parallel to O–O), CH_3OH (parallel to C–O), etc. The x-axis is either selected along one of the bonds to simplify the KEO or as a bisector of a dihedral (book) angle to retain the molecule symmetry and to coincide with the PAS structure at the equilibrium.

These orientational choices are purely geometrical and can always be (at least if chosen sensibly) associated with three conditions. For our example of HCN, the xyz frame oriented parallel to the C–N bond, these three geometric conditions (see Fig. 2.5) are given by

$$\begin{aligned} r_{\mathrm{N},x} - r_{\mathrm{C},x} &= 0, \\ r_{\mathrm{N},y} - r_{\mathrm{C},y} &= 0, \\ r_{\mathrm{H},y} - r_{\mathrm{C},y} &= 0, \end{aligned} \tag{2.18}$$

where the first two lines require that the x and y components of the C–N bond vector vanish while the third condition is to place the C–H bond in the xy plane with the zero y component. Other examples of geometrically defined frames are illustrated below.

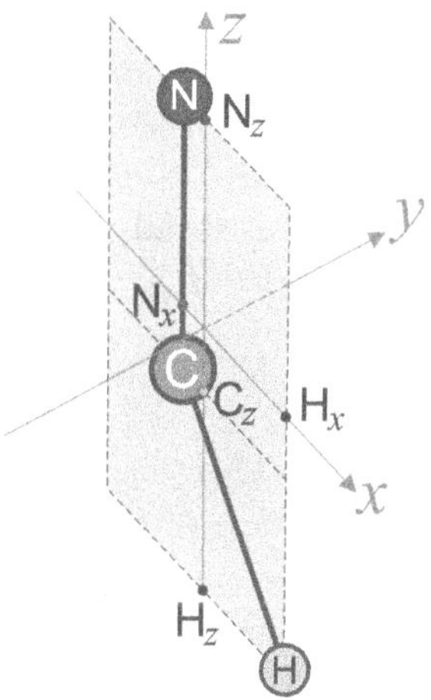

Figure 2.5 A bond xyz frame for HCN with the centre at the centre-of-mass and the z axis parallel to the CN bond. The nuclear Cartesian coordinates are indicated with small circles.

It should be noted that in this example the frame axes are chosen parallel to the molecular bonds but not to coincide with them. The centre of the coordinates should always be at the centre-of-mass which usually does not coincide with the nuclear positions. It is possible, and even often preferable, to orient the xyz frame along the lines connecting the nuclei with the nuclear centre-of-mass (or with the local centre of mass). Such embeddings are called Jacobi and are often associated with the Jacobi-type vibrational coordinates, also defined using connections between the nuclei and centre-of-masses (see below).

2.3.2 Eckart frame

The alternative criterion for the embedding is to seek a better separation between the vibrational and rotational degrees of freedom. This is the motivation behind the so-called Eckart frame (EF) introduced by Eckart (1935).

An EF is defined using the following (three) conditions:

$$\sum_{n=1}^{N} m_n \boldsymbol{r}_n^{\mathrm{e}} \times \boldsymbol{r}_n = 0, \tag{2.19}$$

where $\boldsymbol{r}_n$ is the Cartesian coordinate vector of the nth nucleus in an xyz BF frame and $\boldsymbol{r}_n^{\mathrm{e}}$ is its equilibrium position (the superscript BF is omitted). These conditions are derived from a conjecture that in order to eliminate the Coriolis coupling and to fully separate the rotation and vibrational degrees of freedom in the KEO, one would need to (classically) eliminate the vibrational angular momentum

$$\boldsymbol{L}_{\mathrm{vib}} = \sum_i m_i \boldsymbol{r}_i \times \dot{\boldsymbol{r}}_i,$$

where $\boldsymbol{r}_i$ are the coordinate vector and associated velocity vector of the nucleus i with the mass m_i, respectively. One can show that with only three parameters θ, ϕ and χ at hand it is not possible to fully eliminate the Coriolis coupling from the quantum-mechanical KEO. Therefore, instead, Eckart (1935) suggested to eliminate the following approximate expression for $\boldsymbol{L}_{\mathrm{vib}}$:

$$\boldsymbol{L}_{\mathrm{vib}} \approx \sum_i m_i \boldsymbol{r}_i^{\mathrm{e}} \times \dot{\boldsymbol{r}}_i, \tag{2.20}$$

where $\boldsymbol{r}_i^{\mathrm{e}}$ defines the equilibrium position of the nucleus i, which is a good approximation for $\boldsymbol{L}_{\mathrm{vib}}$ at least for rigid molecules. The xyz frame (i.e. the orientation of xyz relative to XYZ) is then defined to satisfy the conditions

$$\sum_i m_i \boldsymbol{r}_i^{\mathrm{e}} \times \dot{\boldsymbol{r}}_i = 0,$$

which is consistent with the Eckart conditions given in Eq. (2.19) after taking the first derivative with respect to time.

The equilibrium coordinates $\boldsymbol{r}_n^{\mathrm{e}}$ required by the Eckart conditions are usually assumed to be in the (equilibrium) PAS frame, i.e. with the corresponding tensor of moments of inertia diagonal, or mathematically:

$$\sum_{n=1}^{N} m_n r_{n\alpha}^{\mathrm{e}} r_{n\beta}^{\mathrm{e}} = 0, \tag{2.21}$$

for three pairs of $(\alpha, \beta) = (xy,\ zx,\ yz)$. The orientation of the equilibrium PAS frame is fully defined by the equilibrium structure of the molecule via Eq. (2.21).

Three equations (2.19) in conjunction with the centre-of-mass conditions given in Eq. (2.5) and the equilibrium configuration of the molecule are general prerequisites to define the EF or Eckart embedding, and can in principle be used in connection with any arbitrary system of the vibrational coordinates. For a triatomic (planar) molecule, it is relatively straightforward to construct the Eckart orientation. Indeed, if we place the molecule in the xz plane with $r_{iy} = 0$ for any $i = 1, 2, 3$, then two out of three conditions (on x and z) in Eq. (2.19) are satisfied automatically. The third, y, condition requires a simple, single rotation about the y axis (which is orthogonal to the molecular plane) and can be easily solved for. The general case of a non-planar molecule is, however, not so straightforward. The Eckart conditions are written in terms of the Cartesian displacements (so called rectilinear coordinates), while the actual vibrational coordinates can be anything including curvilinear coordinates (e.g. valence), leading to transcendental equations. For this reason, using the rectilinear coordinates (such as normal modes discussed below) is a more natural choice to deal with the Eckart conditions. Moreover, corresponding KEO in the EF is not in the simplest form. Still, the EF is a popular choice regardless of the vibrational coordinates used owing to its efficient separation between the rotational and vibrational degrees of freedom for any types of coordinates. More importantly for the purpose of this book, the Eckart conditions provide a simple general algorithmic formulation of a molecular frame, which makes it efficient for numerical applications for general cases of arbitrary systems.

2.3.3 Principal axes frame

As was pointed out above, a PAS seems to be a natural choice for defining a body-fixed (BF) xyz frame not only for the equilibrium frame given in Eq. (2.21). A general PAS embedding of a vibrating molecule requires that the instantaneous tensor of the moments of inertia is diagonal at any arbitrary configuration of nuclei, i.e.

$$
\begin{aligned}
I_{\alpha,\alpha}^{\mathrm{PAS}} &= \sum_{n=1}^{N} m_n \left[(r_{n\beta}^{\mathrm{PAS}})^2 + (r_{n\gamma}^{\mathrm{PAS}})^2 \right], \\
I_{\beta,\gamma}^{\mathrm{PAS}} &= \sum_{n=1}^{N} m_n r_{n\beta}^{\mathrm{PAS}} r_{n\gamma}^{\mathrm{PAS}} = 0, \quad \beta \neq \gamma,
\end{aligned}
\tag{2.22}
$$

where $\alpha, \beta,\ \gamma$ are x, y, z or their cyclic permutations. In principle, the PAS also provides a simple and general set of rules to define the body axis frame and should be well suited for a general purpose numerical algorithm. However, it is known to suffer from singularities at some (accessible via even usual vibrational motions) geometries and therefore not so useful in practical ro-vibrational applications. These singularities can appear even at non-linear geometries when the moments of inertia become degenerate, which is illustrated in Section 4.1.3 for an XY_2 molecule.

2.4 VIBRATIONAL COORDINATES

Let us assume that a BF frame for our N-atomic molecule has been defined for an arbitrary instantaneous position of the nuclei in space. Now we place an observer at the centre-of-mass of the rotating molecule who watches (measures) the relative motion of the nuclei in this moving frame (see, e.g., Fig. 2.1). Having used six (five for a linear molecule) conditions to define the molecular-fixed coordinate system, we are left with $3N-6$ (or $3N-5$) constraints to choose the internal, 'vibrational' degrees of freedom. In the following we review some popular choices of vibrational coordinates and associated conditions.

2.4.1 Geometrically defined, valence coordinates

Arguably the most intuitive choice of vibrational coordinates is associated with the so-called valence coordinates: the bond lengths, inter-bond angles and dihedral angles (book angles), also commonly referred to as geometrically defined coordinates (GDC) or curvilinear coordinates (for a reason which will become clear later). For example, the internal configuration of a vibrating triatomic ($N = 3$) water molecule (H_2O) can be fully described by three ($3N - 6 = 3$) internal degrees of freedom: bond lengths r_1 and r_2 and an inter-bond angle α, as also shown in Fig. 2.6.

A four-atomic ($N = 4$) molecule hydrogen peroxide (H_2O_2 illustrated in Fig. 2.7) has three bonds with lengths r_{12}, r_{23} and r_{34}; two inter-bond angles α_{123} and α_{234}; and one dihedral (book angle) between two planes O–O–H_1 and O–O–H_2, δ_{1234}.

By construction, these valence coordinates can always be defined in terms of the Cartesian coordinates in any frame. As we need $3N-6$ constraints to specify

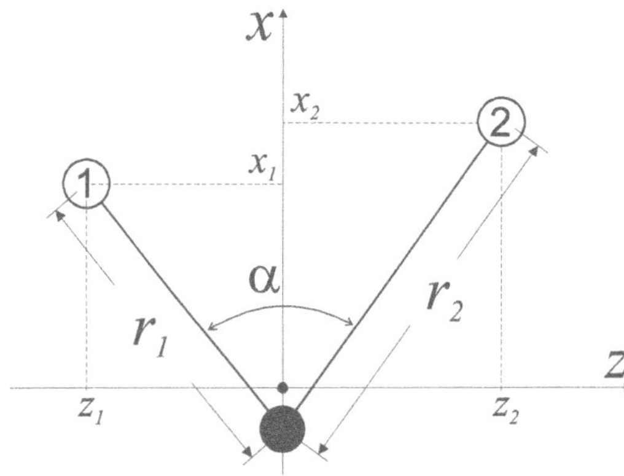

Figure 2.6 An example of geometrically defined (valence) coordinates for a symmetric XY_2 molecule: two bond lengths r_1 and r_2 and a bond angle α.

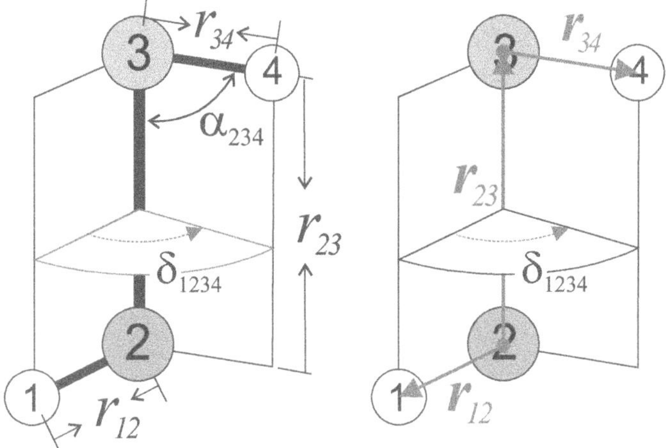

Figure 2.7 Geometrically defined (valence) coordinates for a tetratomic molecule HOOH: three bond lengths r_{12}, r_{23}, r_{34}; two bond angles α_{123} and α_{234}, a dihedral (book) angle δ_{1234}, and associated bond vectors.

the coordinate transformation in Eq. (2.4), the formal definitions of the $3N-6$ valence coordinates in terms of $3N$ Cartesian coordinates $\boldsymbol{r}_n$ give us all the required conditions.

The bond length coordinate r_{ij} is a distance between the nuclei i and j:

$$r_{ij} = \sqrt{(\boldsymbol{r}_i - \boldsymbol{r}_j)^2} = \sqrt{\sum_{\alpha=x,y,z}^{N} (r_{i\alpha} - r_{i\alpha})^2}, \tag{2.23}$$

where $\boldsymbol{r}_i$ is a coordinate vector pointing to the nucleus i from the centre of coordinates. The bond angle α_{ijk} between two bond vectors $\boldsymbol{r}_{ij}$ and $\boldsymbol{r}_{jk}$ (atom j is in the middle; see Fig. 2.7) is defined via their scalar products:

$$\alpha_{ijk} = \arccos\left(\frac{\boldsymbol{r}_{ij}\cdot\boldsymbol{r}_{jk}}{r_{ij}r_{jk}}\right) = \arccos\left(\frac{\sum_{\alpha=x,y,z}(r_{i\alpha}-r_{j\alpha})(r_{k\alpha}-r_{j\alpha})}{r_{ij}r_{jk}}\right). \tag{2.24}$$

A dihedral angle (book angle) δ_{ijkl} can be defined as an angle between two planes containing the bond vectors $\boldsymbol{r}_{ij}$, $\boldsymbol{r}_{jk}$ and $\boldsymbol{r}_{jk}$, $\boldsymbol{r}_{kl}$ with normals $\boldsymbol{a}$ and $\boldsymbol{b}$ (see Fig. 2.8) via the scalar product $\boldsymbol{a}\cdot\boldsymbol{b}$:

$$\delta_{ijkl} = \arccos\left(\frac{\boldsymbol{a}\cdot\boldsymbol{b}}{|\boldsymbol{a}||\boldsymbol{b}|}\right) = \arccos\left(\frac{[\boldsymbol{r}_{ji}\times\boldsymbol{r}_{jk}]\cdot[\boldsymbol{r}_{jk}\times\boldsymbol{r}_{kl}]}{|[\boldsymbol{r}_{ji}\times\boldsymbol{r}_{jk}]||[\boldsymbol{r}_{jk}\times\boldsymbol{r}_{kl}]|}\right). \tag{2.25}$$

The normals $\boldsymbol{a}$ and $\boldsymbol{b}$ are formed as vector products of the corresponding in-plane valence vectors $\boldsymbol{u}$, $\boldsymbol{v}$ and $\boldsymbol{v}$, $\boldsymbol{w}$

$$\begin{aligned} \boldsymbol{a} &= \boldsymbol{u}\times\boldsymbol{v}, \\ \boldsymbol{b} &= \boldsymbol{v}\times\boldsymbol{w}, \end{aligned} \tag{2.26}$$

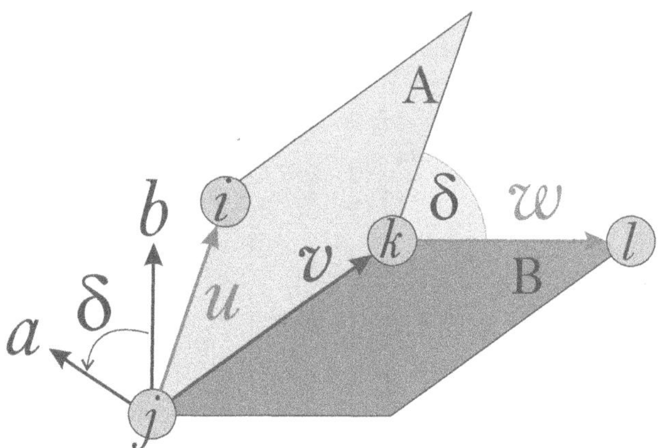

Figure 2.8 Definition of the book (dihedral) angle δ between the planes A and B formed by the three vectors $\boldsymbol{u}$, $\boldsymbol{v}$ and $\boldsymbol{w}$ connecting the four nuclei i, j, k and l. The vectors $\boldsymbol{a}$ and $\boldsymbol{b}$ are the normals to the planes A and B, respectively.

where

$$\begin{aligned} \boldsymbol{u} &\equiv \boldsymbol{r}_{ji} = \boldsymbol{r}_i - \boldsymbol{r}_j, \\ \boldsymbol{v} &\equiv \boldsymbol{r}_{jk} = \boldsymbol{r}_k - \boldsymbol{r}_j, \\ \boldsymbol{w} &\equiv \boldsymbol{r}_{kl} = \boldsymbol{r}_l - \boldsymbol{r}_k. \end{aligned} \tag{2.27}$$

The definition of the dihedral angle depends on the orientation of the cross-products given in Eq. (2.25). For example, the dihedral angle δ_{1234} in Fig. 2.7 is defined using a counter-clockwise direction as follows:

$$\delta_{1234} = \arccos\left(\frac{[\boldsymbol{r}_{21} \times \boldsymbol{r}_{23}] \cdot [\boldsymbol{r}_{23} \times \boldsymbol{r}_{34}]}{|\boldsymbol{r}_{21} \times \boldsymbol{r}_{23}|\,|\boldsymbol{r}_{23} \times \boldsymbol{r}_{34}|}\right).$$

The dihedral angle with the opposite orientation is formulated by inverting the middle vector $\boldsymbol{v}$ to $-\boldsymbol{v}$.

The arccos formulation of δ_{ijkl} in Eq. (2.25) suffers, however, from the limited definition range of the dihedral angle from 0 to 180°. If the full coverage of $[0, 360°]$ is important, it can be supplemented with the equivalent arcsin formulation:

$$\delta_{ijkl} = \arcsin\left(\frac{[[\boldsymbol{r}_{jk} \times \boldsymbol{r}_{kl}] \times [\boldsymbol{r}_{ji} \times \boldsymbol{r}_{jk}]] \cdot \boldsymbol{r}_{kl}}{|\boldsymbol{r}_{ji} \times \boldsymbol{r}_{jk}||\boldsymbol{r}_{jk} \times \boldsymbol{r}_{kl}||\boldsymbol{r}_{kl}|}\right). \tag{2.28}$$

The definition of the valence coordinates is frame independent. The bond lengths r_{ij}, inter-bond angles α_{ijk} and the dihedral angles δ_{ijkl} do not depend on the translational position or orientation of the frame and thus can be equally expressed in terms of the LF frame Cartesian $\boldsymbol{R}_n^{\mathrm{LF}}$ or BF frame $\boldsymbol{r}_n^{\mathrm{BF}}$ coordinates.

For future reference, let us introduce a generalised coordinate vector $\boldsymbol{\xi}_n$ ($n = 1 \ldots 3N-6$) constructed from the $N-1$ bond lengths r_{ij}, $N-2$ bond angles α_{ijk}

and $N-3$ dihedral angles δ_{ijkl}:

$$\boldsymbol{\xi} = \{\underbrace{r_{ij}}_{N-1},\ \underbrace{\alpha_{ijk}}_{N-2},\ \underbrace{\delta_{ijkl}}_{N-3}\}.$$

The relations between the valence and Cartesian coordinates are relatively simple and general and are therefore well suited for numerical applications. However, this is true only in one direction, from $\boldsymbol{R}_n$ to $\boldsymbol{\xi}_i$. The inverse transformation in general does not have a closed, analytic form; it depends on the orientation and position of the coordinate frame and is therefore subject to the additional six constraints introduced above, which complicates the derivation of the KEO for the transformation given in Eq. (2.4). It is possible, however, to formulate the coordinate transformation between $\boldsymbol{R}_n$ and $\boldsymbol{\xi}_i$ as a general (and efficient) numerical algorithm, as will be shown in the following sections.

2.4.2 Z-matrix

A drawback of the valence coordinates, at least for our purposes, is that their definition becomes ambiguous for larger systems as there are different choices to define inter-bond vectors $\boldsymbol{r}_{ij}$. An attempt to generalise this definition uses so-called Z-matrix coordinates and the associated Z-matrix frame. As we will see, for any non-linear N-atomic molecule, one can always choose $3N-6$ independent valence coordinates as $N-1$ bond lengths, $N-2$ inter-bond angles and $N-3$ dihedral angles.

Let us assume a general N atomic molecule, with all atoms numbered from 1 to N. The valence coordinates, bond lengths r_{ij}, inter-bond angles α_{ijk} and dihedral angles δ_{ijkl} as well as a BF xyz frame are defined as part of the construction of the Z-matrix as follows. The first three steps are illustrated in Fig. 2.9.

- We start by centring the Z-matrix frame at nucleus 1, i.e. at $x_1^{\rm ZM} = y_1^{\rm ZM} = z_1^{\rm ZM} = 0$.
- The z-axis is oriented from nucleus 1 to nucleus 2. Here we introduce the first valence coordinate, the distance r_{12} (bond length) between atoms 1 and 2. The body-frame position of atom 2 is at $x_2 = 0, y_2 = 0$ and $z_2 = \pm r_{12}$, where r_{12} is the length of the 1–2 bond.
- (Choice 1: linking 3 to 1, see the left display of Fig. 2.9) The xz plane is then oriented to contain the inter-nuclear bond 1–3 at distance r_{13} from nucleus 1, making the angle α_{213} with the bond 1–2 and Cartesian coordinates given by

$$x_3^{\rm ZM} = r_{13}\sin(\alpha_{213}), \tag{2.29}$$
$$y_3^{\rm ZM} = 0, \tag{2.30}$$
$$z_3^{\rm ZM} = r_{13}\cos(\alpha_{213}). \tag{2.31}$$

  This gives two more generalised valence coordinates $\xi_2 = r_{13}$ and $\xi_3 = \alpha_{213}$.

- (Choice 2: linking 3 to 2, see the right display of Fig. 2.9) Alternatively, the xz plane is oriented to contain the inter-nuclear bond 2–3 at distance r_{23} from nucleus 2, making the angle α_{123} with the bond 1–2 and Cartesian coordinates given by

$$x_3^{\mathrm{ZM}} = r_{23}\sin(\alpha_{123}), \tag{2.32}$$
$$y_3^{\mathrm{ZM}} = 0, \tag{2.33}$$
$$z_3^{\mathrm{ZM}} = r_{12} - r_{23}\cos(\alpha_{123}). \tag{2.34}$$

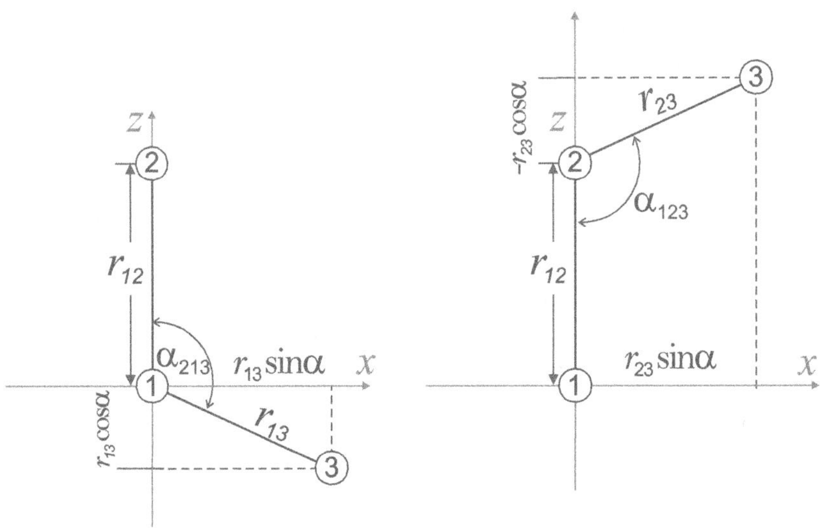

Figure 2.9 Two orientations of the Z-matrix frame for a symmetric XY_2 molecule.

This step completes the definition of the Z-matrix frame, but the Z-matrix formalism continues with setting up the Cartesian positions of the remaining nuclei.

- For any following atom i ($i = 4, \ldots N$), see Fig. 2.8, three generalised valence coordinates

$$\xi_{i-1} = r_{ij}, \tag{2.35}$$
$$\xi_{i-2} = \alpha_{ijk}, \tag{2.36}$$
$$\xi_{i-3} = \delta_{ijkl} \tag{2.37}$$

are defined as follows. Three reference atoms j, k and l are selected from already specified positions. A polar spherical vector $\boldsymbol{r}_{ij} = (r_{ij}, \alpha_{ijk}, \delta_{ijkl})$ is defined to point from atom j to atom i

$$\boldsymbol{r}_i^{\mathrm{ZM}} = \boldsymbol{r}_j^{\mathrm{ZM}} + \boldsymbol{r}_{ij}$$

with

$$\begin{aligned}\boldsymbol{r}_{ij} \;=\; & r_{ij}\,(\boldsymbol{e}_{jk}\cos\alpha_{ijk} + \\ & \boldsymbol{e}_{\perp}\sin\alpha_{ijk}\,\cos\delta_{ijkl} + \boldsymbol{e}_{jkl}\sin\alpha_{ijk}\,\sin\delta_{ijkl}),\end{aligned} \tag{2.38}$$

where $\boldsymbol{r}_j$ is the position vector of the atom j and the unit vectors $\boldsymbol{e}_{jk}$, $\boldsymbol{e}_{\perp}$ and $\boldsymbol{e}_{jkl}$ are given by

$$\boldsymbol{e}_{jk} = \frac{\boldsymbol{r}_{jk}}{|\boldsymbol{r}_{jk}|}, \tag{2.39}$$

$$\boldsymbol{e}_{jkl} = \frac{\boldsymbol{r}_{jk} \times \boldsymbol{r}_{jl}}{|\boldsymbol{r}_{jk}|\,|\boldsymbol{r}_{kl}|}, \tag{2.40}$$

$$\boldsymbol{e}_{\perp} = \boldsymbol{e}_{jk} \times \boldsymbol{e}_{jkl}. \tag{2.41}$$

The direction of the dihedral angle δ_{ijkl} needs to be specified: positive (counter-clockwise) or negative (clockwise) depending on the direction of the $\boldsymbol{v}$ vector in Fig. 2.8.

For example, the Z-matrix system for an XY_2 molecule from the left display of Fig. 2.9 is given by

$$\begin{aligned} x_{\mathrm{O}}^{\mathrm{ZM}} &= 0, & x_{\mathrm{H}_1}^{\mathrm{ZM}} &= 0, & x_{\mathrm{H}_2}^{\mathrm{ZM}} &= r_2 \sin\alpha, \\ y_{\mathrm{O}}^{\mathrm{ZM}} &= 0, & y_{\mathrm{H}_1}^{\mathrm{ZM}} &= 0, & y_{\mathrm{H}_2}^{\mathrm{ZM}} &= 0, \\ z_{\mathrm{O}}^{\mathrm{ZM}} &= 0, & z_{\mathrm{H}_1}^{\mathrm{ZM}} &= r_1, & z_{\mathrm{H}_2}^{\mathrm{ZM}} &= r_2 \cos\alpha. \end{aligned}$$

The definition of the Z-matrix is easily formalised and is therefore very practical for numerical applications. The Z-matrix formalism is an integral part of any electronic structure code. Although it is not necessarily efficient as a molecular embedding, for a non-symmetric molecule it can provide a physically intuitive orientation. Such an example is the tetratomic HFCO shown with its Z-matrix embedding in Fig. 2.10. It can also suffer from the redundancy of the coordinates as, e.g., in the case of CH_4; see Section 2.9. In general, however, the Z-matrix can serve as a useful intermediate frame, with the advantage that the definition of different properties can be automated and then transformed (rotated) to the desired frame.

In order to turn the Z-matrix frame into a formally defined BF frame xyz, the Z-matrix coordinate centre (defined at atom 1) needs to be shifted to the centre-of-mass with the corresponding shifts of the Cartesian coordinate vectors $\boldsymbol{r}_i^{\mathrm{ZM}}$:

$$\boldsymbol{r}_n \equiv \boldsymbol{r}_n^{\mathrm{BF}} = \boldsymbol{r}_n^{\mathrm{ZM}} - \boldsymbol{r}^{\mathrm{CM}},$$

where the CM vector $\boldsymbol{r}^{\mathrm{CM}}$ is given by

$$\boldsymbol{r}^{\mathrm{CM}} = \sum_n \frac{m_n}{M} \boldsymbol{r}_n^{\mathrm{ZM}}.$$

It is useful to point out that the expressions above fully define the Cartesian coordinates of each nuclei in the Z-matrix for any instantaneous configuration in terms of the vibrational coordinates $\boldsymbol{\xi}$. That is, for any arbitrary numerical set of $\xi_1 \ldots \xi_M$ ($M = 3N - 6$), these expressions define the corresponding values of the Cartesian coordinates. This is useful for the numerical evaluation of the KEO at any instantaneous nuclear configuration.

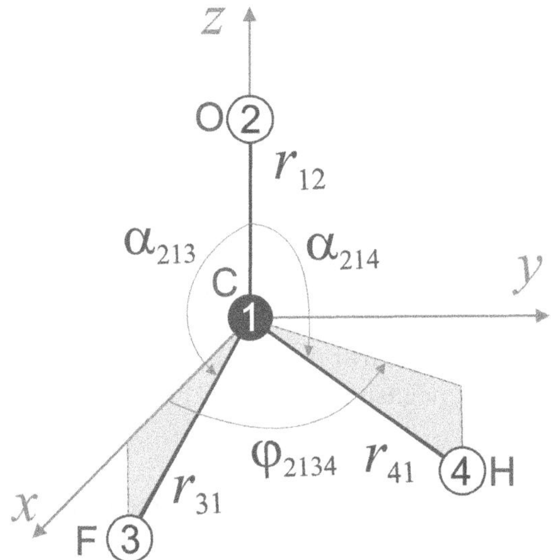

Figure 2.10 Example of Z-matrix coordinates and associated embedding for the HFCO molecule.

As will be shown in the next chapter, construction of the KEO in the molecular frame representation requires derivatives of the Cartesian coordinates in a given frame with respect to the $3N - 6$ internal coordinates. The Z-matrix formulation allows for analytic evaluation of the derivatives of $\boldsymbol{r}_{ik}^{\mathrm{ZM}}$ with respect to the valence coordinates r_{ij}, α_{ijk} and δ_{ijkl} via Eq. (2.38). It should also be noted that a full derivation of the KEO for the coordinate system associated with a body-frame system requires such derivatives up to the third order.

2.4.3 Jacobi frames

The Jacobi coordinates are defined sequentially as vectors (called Jacobi vectors) connecting centres-of-mass of the previously considered particles to new particles in different configurations (Jepsen and Hirschfelder, 1959).

An example of a Jacobi embedding with Jacobi coordinates r_1, r_2 and α for an XY_2 molecule is shown in Fig. 2.11, where r_1 and r_2 are distances from the nuclear centre-of-mass and the two nuclei Y_1 and Y_2 and α is the angle between them. The corresponding vectors $\boldsymbol{r}_1$ and $\boldsymbol{r}_2$ represent Jacobi vectors. Here the xyz coordinate centre is at the centre-of-mass with the x axis chosen to bisect the angle α, the z axis lying in the plane of the molecule and the y axis perpendicular to the plane (not shown) in the right-handed sense. Jacobi frames/coordinates lead to simpler, more compact KEOs and are often referred to as orthogonal coordinates.

For a general polyatomic molecule, standardised Jacobi's treatment (Aquilanti and Cavalli, 1986) can be defined as follows: we start by defining a vector $\boldsymbol{r}_1$ that connects the atoms 1 and 2. The other $N - 2$ vectors $\boldsymbol{r}_i$ are directed from the centre-of-mass of the $i - 1$ atoms to the ith atom (or *vice versa*). Analogously

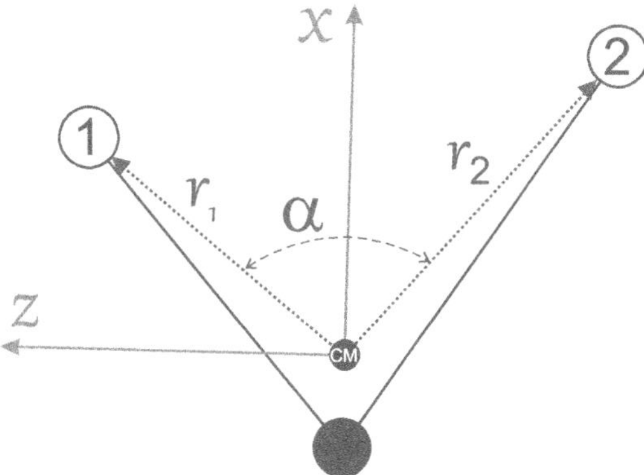

Figure 2.11 An example of a Jacobi frame with three Jacobi coordinates for a symmetric XY_2 molecule.

to the Z-matrix procedure, each vector $\boldsymbol{r}_i$ introduces internal coordinates r_i ($i = 1, \ldots, N-1$), θ_i ($i = 2, \ldots, N-1$) and ϕ_i ($i = 3, \ldots, N-1$). A BF frame xyz can be also defined according to the Z-matrix scheme described above. Figure 2.12 (left) illustrates Jacobi vectors $\boldsymbol{r}_1$, $\boldsymbol{r}_2$ and $\boldsymbol{r}_3$ for a general tetratomic molecule ABCD, together with corresponding six Jacobi coordinates r_1, r_2, θ_2, r_3, θ_3 and ϕ_3.

Depending on the type of the system, different alternative definitions of the Jacobi vectors can be introduced by dividing atoms into subgroups and using the above procedure to build Jacobi vectors within each set. This is illustrated for an ABCD molecule in the right-hand side of Fig. 2.12 using a popular diatom-diatom Jacobi scheme.

2.4.4 Derivatives of the geometrically defined coordinates with respect to the Cartesian coordinates

While we are on the topic of evaluating derivatives of the coordinates, let us also consider the derivatives of the valence coordinates ξ_i ($i = 1, \ldots, M$) with respect to the Cartesian coordinates $r_{n\alpha}$ ($n = 1, \ldots, N$ and $\alpha = x, y, z$).

Using the definition of the valence coordinates in Eqs. (2.23–2.25), for derivatives of the valence bonds and angles with respect to the Cartesian coordinates, we obtain

$$\frac{\partial r_{ij}}{\partial r_{i\alpha}} = -\frac{\partial r_{ij}}{\partial r_{j\alpha}} = \frac{r_{i\alpha} - r_{j\alpha}}{|\boldsymbol{r}_{ij}|}, \tag{2.42}$$

$$\frac{\partial \theta_{ijk}}{\partial r_{l\alpha}} = -\frac{\partial u_{ijk}}{\partial r_{j\alpha}} \frac{1}{\sin\theta_{ijk}} \tag{2.43}$$

where we introduced a shorthand notation

$$u_{ijk} \equiv \cos\theta_{ijk}$$

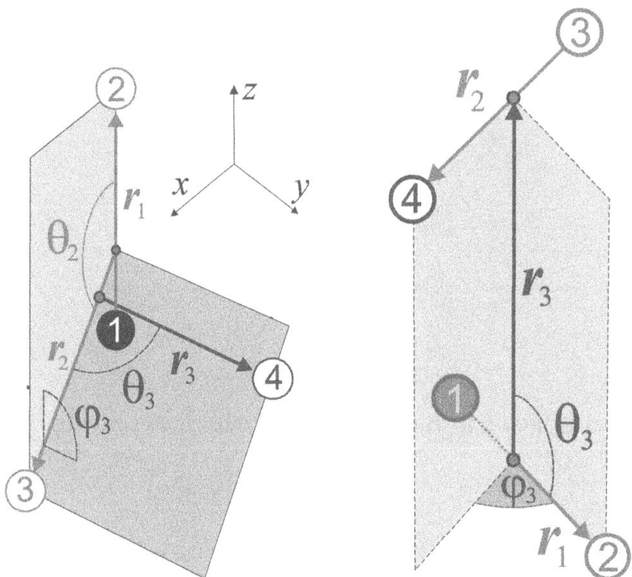

Figure 2.12 An example of Jacobi frame coordinates for an ABCD molecule, r_1, r_2, θ_2, r_3, θ_3 and ϕ_3: atom-triatom (left) and diatom-diatom (right).

and

$$\frac{\partial u_{ijk}}{\partial r_{i\alpha}} = \frac{r_{k\alpha} - r_{j\alpha}}{|\boldsymbol{r}_{ij}||\boldsymbol{r}_{jk}|} - \frac{r_{i\alpha} - r_{j\alpha}}{|\boldsymbol{r}_{ij}|^2} u_{ijk}, \tag{2.44}$$

$$\frac{\partial u_{ijk}}{\partial r_{j\alpha}} = -\frac{r_{i\alpha} - r_{j\alpha}}{|\boldsymbol{r}_{ij}||\boldsymbol{r}_{jk}|}\left(1 - u_{ijk}\frac{|\boldsymbol{r}_{kj}|}{|\boldsymbol{r}_{ij}|}\right) - \frac{r_{k\alpha} - r_{j\alpha}}{|\boldsymbol{r}_{ij}||\boldsymbol{r}_{jk}|}\left(1 - u_{ijk}\frac{|\boldsymbol{r}_{ij}|}{|\boldsymbol{r}_{kj}|}\right). \tag{2.45}$$

The expressions for the derivatives of the dihedral angle are more complicated:

$$\frac{\partial \delta_{ijkl}}{\partial \boldsymbol{r}_i} = \frac{|\boldsymbol{r}_{kj}|}{|\boldsymbol{a}|^2}\boldsymbol{a}, \tag{2.46}$$

$$\frac{\partial \delta_{ijkl}}{\partial \boldsymbol{r}_l} = -\frac{|\boldsymbol{r}_{nk}|}{|\boldsymbol{b}|^2}\boldsymbol{b}, \tag{2.47}$$

$$\frac{\partial \delta_{ijkl}}{\partial \boldsymbol{r}_j} = \left(\frac{(\boldsymbol{u}\cdot\boldsymbol{v})}{|\boldsymbol{v}|^2} - 1\right)\frac{\partial \delta_{ijkl}}{\partial \boldsymbol{r}_i} - \frac{(\boldsymbol{w}\cdot\boldsymbol{v})}{|\boldsymbol{v}|^2}\frac{\partial \delta_{ijkl}}{\partial \boldsymbol{r}_l}, \tag{2.48}$$

$$\frac{\partial \delta_{ijkl}}{\partial \boldsymbol{r}_k} = \left(\frac{(\boldsymbol{v}\cdot\boldsymbol{w})}{|\boldsymbol{v}|^2} - 1\right)\frac{\partial \delta_{ijkl}}{\partial \boldsymbol{r}_l} - \frac{(\boldsymbol{u}\cdot\boldsymbol{v})}{|\boldsymbol{v}|^2}\frac{\partial \delta_{ijkl}}{\partial \boldsymbol{r}_i}. \tag{2.49}$$

where the bond vectors $\boldsymbol{u}$, $\boldsymbol{v}$ and $\boldsymbol{w}$ are as given in Eq. (2.27), while the normals $\boldsymbol{a}$ and $\boldsymbol{b}$ are defined in Eq. (2.26). As a reminder, the definition of the dihedral angle δ in Eq. (2.25) has the compact form:

$$\delta = \arccos\left(\frac{\boldsymbol{a}\cdot\boldsymbol{b}}{|\boldsymbol{a}||\boldsymbol{b}|}\right).$$

It should also be mentioned that in practical applications we tend to use displacements of ξ_i from their equilibrium positions. This will be taken into account when the above expressions are utilised for constructing the KEO.

2.5 LINEARISED COORDINATES

Valence or any other geometrically defined coordinates are useful as they provide a physically or chemically intuitive definition of the vibrational degrees of freedom. Valence coordinates are well suited for the description of molecular inter-nuclear forces and potential energy functions. Here we define the so-called rectilinear coordinates which are especially suited for use with the EF.

The rectilinear (vibrational) coordinates are defined as linear combinations of Cartesian coordinates and their displacements from the equilibrium configuration:

$$\xi_i^{\ell} \equiv \sum_{n,\alpha} B_{i,n\alpha}(r_{n\alpha} - r_{n\alpha}^{\mathrm{e}}).$$

Let us consider the transformation from the LF frame XYZ to a body frame xyz in Eq. (2.4) and express the body-frame position vectors $\boldsymbol{r}_n$ via displacements from the equilibrium configuration as follows:

$$R_{i,A} = R_A^{\mathrm{CM}} + \sum_{\alpha} S_{A,\alpha}(\theta,\phi,\chi)\left[r_{i,\alpha}^{\mathrm{e}} + \Delta r_{i,\alpha}(\boldsymbol{\xi})\right], \tag{2.50}$$

where $i = 1,\ldots,N$; $A = X,Y,Z$; $\alpha = x,y,z$. In this equation, θ,ϕ and χ are the Euler angles and $\boldsymbol{r}_i^{\mathrm{e}}$ are the BF equilibrium positions of the rotating molecule associated with a body frame xyz. Let us first represent the Cartesian displacements $\Delta\boldsymbol{r}_i$ as linear combinations of $3N-6$ rectilinear coordinates ξ_λ ($\lambda = 1\ldots 3N-6$):

$$r_{i\alpha} = r_{i\alpha}^{\mathrm{e}} + \Delta_{i,\alpha} = r_{i\alpha}^{\mathrm{e}} + \sum_{\lambda} A_{i\alpha,\lambda}\,\xi_\lambda, \tag{2.51}$$

where $A_{i\alpha,\lambda}$ are linear transformational elements ($N\times 3\times(3N-6)$ in total). The two most common examples of rectilinear coordinates are the ‘normal’ and ‘linearised’ coordinates.

The so-called linearised coordinates ξ_λ^{ℓ} are defined as linearised versions of the valence coordinates r_{ij}, α_{ijk} and δ_{ijkl} by expanding Eqs. (2.23–2.25) into Taylor series in terms of $\Delta r_{i\alpha} = r_{i\alpha} - r_{n\alpha}^{\mathrm{e}}$ and truncating after 1st order.

As an example, let us consider an XY_2 molecule shown in Fig. 2.13, with two inter-nuclear bonds of lengths $r_1 = r_{13}$ and $r_2 = r_{23}$ and an associated angle between them $\alpha = \alpha_{213}$. A Taylor expansion of r_i ($i = 1,2$) in Eq. (2.23) around

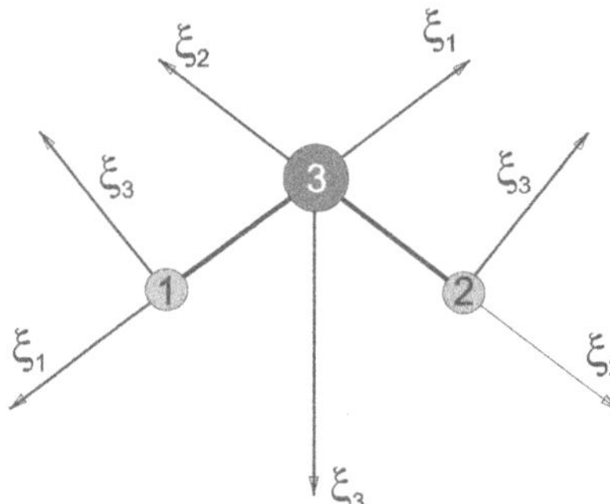

Figure 2.13 Illustration of linearised coordinates ξ_i^ℓ for an XY_2 molecule (to scale), plotted as the corresponding $(B_{i,nx}, B_{i,ny}, B_{i,nx})^T$ vectors for each atom.

their equilibrium values is given by (up to 1st order):

$$r_i - r_{\rm e} = \frac{r^{\rm e}_{i,x} - r^{\rm e}_{3,x}}{r_{\rm e}}(\Delta r_{i,x} - \Delta r_{3,x}) + \frac{r^{\rm e}_{i,z} - r^{\rm e}_{3,z}}{r_{\rm e}}(\Delta r_{i,z} - \Delta r_{3,z}) + O(\Delta^2),$$

or

$$r_i - r_{\rm e} = \sum_{n,\alpha} B_{i,n\alpha}\Delta r_{n,\alpha} + O(\Delta^2), \tag{2.52}$$

with $B_{i,n\alpha}$ as gradients of r_i at equilibrium.

We now define the two stretching linearised coordinates as a linear part of the Taylor expanded $r_i - r_{\rm e}$

$$\xi_i^\ell \equiv \sum_{n,\alpha} B_{i,n\alpha}\Delta r_{n,\alpha} \tag{2.53}$$

with

$$B_{i,i\alpha} = \frac{r^{\rm e}_{i,\alpha} - r^{\rm e}_{3,\alpha}}{r_{\rm e}} \quad \text{and} \quad B_{i,3\alpha} = -B_{i,i\alpha}$$

for $i = 1, 2$ and $\alpha = x, z$. The bond length expansion is then simply

$$r_i = r_{\rm e} + \xi_i^\ell + O(\Delta^2).$$

Using a similar approach for the bending coordinate α, we obtain

$$\alpha - \alpha_{\rm e} = \sum_{n,\alpha} B_{3,n\alpha}\Delta r_{n,\alpha} + O(\Delta^2), \tag{2.54}$$

where the non-zero values of $B_{3,n\alpha}$ defining the bending linearised coordinate as

$$\xi_3^\ell = \sum_{n,\alpha} B_{3,n\alpha}\Delta r_{n,\alpha}$$

are given by

$$B_{3,1\alpha} = -\frac{1}{r_{\mathrm{e}}^2 \sin\alpha_{\mathrm{e}}} \left[(r_{2,\alpha}^{\mathrm{e}} - r_{3,\alpha}^{\mathrm{e}}) - (r_{1,\alpha}^{\mathrm{e}} - r_{3,\alpha}^{\mathrm{e}}) \cos\alpha_{\mathrm{e}}\right], \tag{2.55}$$

$$B_{3,2\alpha} = -\frac{1}{r_{\mathrm{e}}^2 \sin\alpha_{\mathrm{e}}} \left[(r_{1,\alpha}^{\mathrm{e}} - r_{3,\alpha}^{\mathrm{e}}) - (r_{2,\alpha}^{\mathrm{e}} - r_{3,\alpha}^{\mathrm{e}}) \cos\alpha_{\mathrm{e}}\right], \tag{2.56}$$

$$B_{3,3\alpha} = -\frac{1}{r_{\mathrm{e}}^3 \sin\alpha_{\mathrm{e}}} \left[(r_{1,\alpha}^{\mathrm{e}} - r_{3,\alpha}^{\mathrm{e}})(r_2^{\mathrm{e}} \cos\alpha_{\mathrm{e}} - r_1^{\mathrm{e}}) + \right. \tag{2.57}$$

$$\left. + (r_{2,\alpha}^{\mathrm{e}} - r_{3,\alpha}^{\mathrm{e}})(r_1^{\mathrm{e}} \cos\alpha_{\mathrm{e}} - r_2^{\mathrm{e}})\right], \tag{2.58}$$

for $\alpha = x, z$.

In general, a set of $3N - 6$ linearised coordinates ξ_n^ℓ is defined as linear combinations of $\Delta r_{i\alpha}$ and coincides with the set of $3N - 6$ valence coordinates ξ_λ in the linear approximation:

$$\xi_\lambda^\ell = \sum_{i=1}^{N} \sum_{\alpha=x,y,z} B_{\lambda,i\alpha} \Delta r_{i\alpha}; \quad \lambda = 1, 2, \ldots, 3N - 6, \tag{2.59}$$

where

$$B_{\lambda,i\alpha} = \left.\frac{\partial \xi_\lambda}{\partial \Delta r_{i\alpha}}\right|_{\mathrm{e}} = \frac{\partial \xi_\lambda^\ell}{\partial \Delta r_{i\alpha}}. \tag{2.60}$$

Hence, the $\boldsymbol{B}$-matrix, with elements $B_{\lambda,i\alpha}$, can be obtained from purely geometrical considerations by differentiating the valence coordinates with respect to $\Delta r_{i\alpha}$ and evaluating them at equilibrium.

The structural coefficients $A_{i\alpha,n}$ defining the coordinate transformation in Eq. (2.50) via the linearised form in Eq. (2.51) can be obtained by inverting the linear form of Eq. (2.53). Using the chain rules for $\partial\xi_\lambda^\ell / \partial\xi_\mu^\ell$

$$\frac{\partial \xi_\lambda^\ell}{\partial \xi_\mu^\ell} = \sum_{i,\alpha} \frac{\partial \xi_\lambda^\ell}{\partial \Delta r_{i\alpha}} \frac{\partial \Delta r_{i\alpha}}{\partial \xi_\mu^\ell},$$

with the definition of the matrices $\boldsymbol{A}$ and $\boldsymbol{B}$ in Eqs. (2.51) and (2.60), respectively, we obtain an $M \times M$ ($M = 3N - 6$) orthogonality relation between $\boldsymbol{A}$ and $\boldsymbol{B}$:

$$\sum_{i=1}^{N} \sum_{\alpha=x,y,z} B_{\lambda,i\alpha} A_{i\alpha,\mu} = \delta_{\lambda,\mu}. \tag{2.61}$$

The relation in Eq. (2.61) has to be supplemented by the translation (centre-of-mass) and rotational constrains introduced above. The centre-of-mass conditions in Eq. (2.68) for the coordinates $r_{i\alpha}$ in Eq. (2.51) give additional $3 \times M$ equations for the elements of the $\boldsymbol{A}$-matrix

$$\sum_{i=1}^{N} m_i A_{i\alpha,n} = 0, \quad \alpha = x, y, z. \tag{2.62}$$

For the rotational part, it is beneficial to assume the Eckart conditions in Eq. (2.19) with the equilibrium frame taken as a PAS as given in Eq. (2.21). In turn, this gives $3 \times M$ more equations for $A_{i\alpha,n}$

$$\sum_{i=1}^{N} \sum_{\beta,\gamma=x,y,z} m_i \left(\epsilon_{\alpha\beta\gamma} A_{i\beta,n} r_{i\gamma}^{\mathrm{e}}\right) = 0, \quad \alpha = x, y, z. \tag{2.63}$$

Here $\epsilon_{\alpha\beta\gamma}$ is the fully antisymmetric Levi-Civita tensor and $\alpha, \beta, \gamma = x, y, z$.

Equations (2.61–2.63) are linear in $A_{i\alpha,n}$ and can be generalised to the matrix equation

$$\boldsymbol{b}\boldsymbol{A} = \mathbb{1}$$

with the following $3N \times 3N$ matrix elements of $\boldsymbol{b}$

$$b_{\alpha,n\beta} = m_n \delta_{\alpha,\beta}, \quad \alpha = x, y, z \quad (1-3), \tag{2.64}$$

$$b_{\alpha',n\beta} = m_n \sum_{\gamma} \epsilon_{\alpha'\beta\gamma} r_{n,\gamma}^{\mathrm{e}}, \quad \alpha' = x, y, z \quad (4-6), \tag{2.65}$$

$$b_{\lambda,n\beta} = B_{\lambda,n,\beta}, \quad \lambda = 1, \ldots, M \quad (7 \ldots 3N-6), \tag{2.66}$$

where $n = 1, \ldots, N$; $\beta, \gamma = x, y, z$; and $M = 3N - 6$. Thus, the elements $A_{i\alpha,n}$ $(n = 3N)$ can be obtained as an inverse of the matrix $\boldsymbol{b}$, $\boldsymbol{A} = \boldsymbol{b}^{-1}$, where $\boldsymbol{A}$ is an extended $3N \times 3N$ matrix. It is important to note that all structural parameters defining $\boldsymbol{b}$ as well as the inversion procedure to obtain $\boldsymbol{A}$ can be formulated purely numerically for a general case of an arbitrary system.

In the case of our XY_2 example, the $\boldsymbol{b}$ matrix can in principle be inverted analytically; however, even in this it is somewhat complicated. A numerical solution of the linear set of equations in Eqs. (2.64, 2.65) is, however, trivial, at least away from the singularity at linear geometry for $\alpha = 180°$.

As an illustration, Fig. 2.13 shows the linearised modes ξ_1^{ℓ}, ξ_2^{ℓ} and ξ_3^{ℓ} from Eq. (2.59) for the triatomic XY_2 case as the corresponding vectors $(B_{i,nx}, B_{i,ny}, B_{i,nx})^T$ $(i = 1, 2, 3)$ for each atom. It demonstrates that the linearised stretching modes follow the vibrations of the individual bonds 1–3 and 2–3 and are exactly parallel to them. The bending vectors of the hydrogen atoms are orthogonal to the bonds with a relatively large bending component of $B_{3,3z}$ of the central oxygen atom 3 corresponding to tiny Cartesian displacements of the heavier nucleus.

It is instructive to visualise the effect of the molecular vibration on the orientation of the molecular frame. Figure 2.14 shows how the EF reacts to a displacement of nuclei from equilibrium for two cases, H_2O and O_3 molecules. In the case of H_2O, where the centre-of-mass is relatively close to the oxygen atom, the displacement of the nuclei does not affect the orientation of the molecular bonds significantly, and therefore this system appears to be more resilient to the choice of the frame. The geometrically defined, bisector frame in Fig. 2.3 should be relatively close to the EF. In the case of O_3, however, where the centre-of-mass is significantly different from

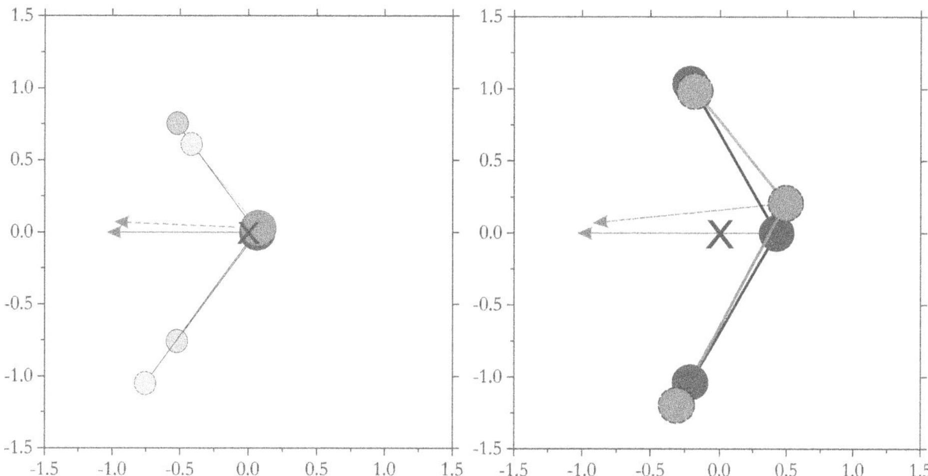

Figure 2.14 Orientational effects upon vibrational motion in H_2O and O_3 assuming the Eckart BF frame and linearised modes shown in Fig. 2.13. The cross indicates the molecular centre-of-mass of the nuclei. An exaggerated displacement is used for illustration purpose.

the position of the central oxygen atom, the vibrational motion is strongly coupled to the rotation affecting the orientational dynamics of the system, and therefore the effect of the frame selection on the solution of the ro-vibrational, nuclear motion problem is expected to be more pronounced.

2.5.1 Normal modes

A famous version of the rectilinear coordinates is the so-called normal modes. Normal coordinates earned their popularity for their formulation as orthogonal coordinates, robustness for building KEO, generality and their direct association with harmonic oscillators, which plays an important role in the description of nuclear motions. If the linearised coordinates are aimed at a physically intuitive representation, the purpose of the normal modes is to make the construction of the KEO as simple as possible, while also preserving the degree of inter-mode separation, at least in the quadratic, harmonic approximation.

The rectilinear normal modes can be defined as part of the coordinate transformation $\boldsymbol{R}_n$ to the BF frame xyz via Eq. (2.50), where the $3N$ Cartesian displacement $\Delta \boldsymbol{r}_n$ are represented as linear combinations of the $3N-6$ coordinates q_λ ($\lambda = 1 \ldots M$, $M = 3N-6$) as follows:

$$\Delta \boldsymbol{r}_n = \sum_{\lambda=1}^{M} \frac{\boldsymbol{l}_{n,\lambda}}{\sqrt{m_n}} q_\lambda. \tag{2.67}$$

The mass-dependent elements of the transformational matrix $\boldsymbol{l}$ relate to the mass-independent form of $\boldsymbol{A}$ by

$$A_{n\alpha,\lambda} = \frac{l_{n\alpha,\lambda}}{\sqrt{m_n}} \quad \alpha = x, y, z.$$

As above, the xyz equilibrium structure $\boldsymbol{r}_n^{\mathrm{e}}$ satisfies the centre-of-mass condition

$$\sum_{i=1}^{N} m_i\, r_{i\alpha}^{\mathrm{e}} = 0 \tag{2.68}$$

(because the xyz axis system has its origin at the nuclear centre-of-mass) and the xyz axis system is traditionally chosen as the PAS at equilibrium:

$$\sum_{i=1}^{N} m_i\, r_{i\alpha}^{\mathrm{e}} r_{i\beta}^{\mathrm{e}} = 0, \quad \alpha \neq \beta. \tag{2.69}$$

The quantities $r_{i\alpha}^{\mathrm{e}}$ can be defined using equilibrium structural parameters of the molecule, such as the bond lengths r_i^{e}, inter-bond $\alpha_{ijk}^{\mathrm{e}}$ and dihedral $\delta_{ijkl}^{\mathrm{e}}$ angles.

The normal-mode matrix elements $\boldsymbol{l}_{n,\lambda}$ from Eq. (2.67) are constructed by satisfying the following set of conditions.

1. The centre-of-mass conditions in Eq. (2.6) lead to the following $3 \times M$ equations:

$$\sum_n \sqrt{m_n} \boldsymbol{l}_{n,\lambda} = 0 \quad \text{or} \quad \sum_n \sqrt{m_n} l_{n\alpha,\lambda} = 0, \quad \alpha = x, y, z. \tag{2.70}$$

2. The three Eckart conditions in Eq. (2.19) correspond to the $3 \times M$ Eckart equations for $l_{n\alpha,\lambda}$ as follows:

$$\sum_n \sqrt{m_n} \boldsymbol{r}_n^{\mathrm{e}} \times \boldsymbol{l}_{n,\lambda} = 0 \quad \text{or} \quad \sum_{\beta,\gamma} \epsilon_{\alpha\beta\gamma} \sum_n \sqrt{m_n} r_{n\beta}^{\mathrm{e}} l_{n\gamma,\lambda} = 0. \tag{2.71}$$

   Here $\epsilon_{\alpha\beta\gamma}$ is the fully antisymmetric tensor again and $\alpha, \beta, \gamma = x, y, z$.

3. The ortho-normality condition is used to define the vibrational coordinates q_λ as orthogonal normal modes and is given by the $M(M-1)/2$ equations

$$\sum_{n,\alpha} l_{n\alpha,\lambda} l_{n\alpha,\lambda'} = \delta_{\lambda,\lambda'}. \tag{2.72}$$

4. The last but not least condition is for the potential energy function $V(\boldsymbol{q})$ to be quadratic and diagonal in the normal-mode coordinates $\boldsymbol{q}$ as given by

$$\frac{\partial^2 V(\boldsymbol{q})}{\partial q_n q_m} = 0, \quad n \neq m. \tag{2.73}$$

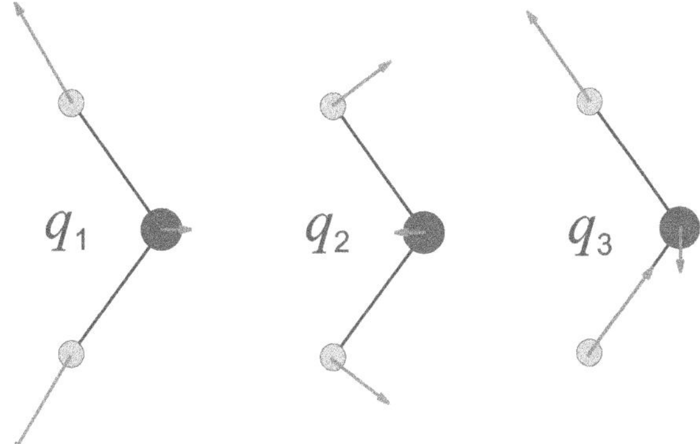

Figure 2.15 Normal-mode coordinates defined as structural vectors $\sqrt{m_n}\,\boldsymbol{l}_{n,\lambda}$ for an XY_2-type molecule H_2O (to scale). Here $n = 1\ldots3$ and $\lambda = 1,2,3$ are the atomic number and vibrational mode indexes, respectively.

Thanks to the orthogonality of $\boldsymbol{l}_{n,\lambda}$ in Eq. (2.72), the inverse transformation to Eq. (2.67) (from $\boldsymbol{r}_n$ to $\boldsymbol{q}$) is directly defined using the same matrix $\boldsymbol{l}$:

$$q_\lambda = \sum_{n,\alpha} \sqrt{m_n} l_{n\alpha,\lambda} \Delta r_{n\alpha}. \tag{2.74}$$

As an example, three normal modes q_1, q_2 and q_3 of a triatomic molecule H_2O are illustrated in Fig. 2.15 by plotting their orthogonal vectors $(l_{\lambda,nx}, l_{\lambda,ny}, l_{\lambda,nz})^T$ ($\lambda = 1,2,3$ and $n = 1,2,3$) and can be compared to the linearised modes in Fig. 2.13. Note that the stretching modes q_1 and q_3 do not exactly follow the molecular bonds 1–3 and 2–3.

As orthogonal vectors, the normal modes automatically transform as irreducible representations of a symmetry group the molecule belongs to. For the $\mathcal{C}_{2\mathrm{v}}(\mathrm{M})$ molecular symmetry group of XY_2, these are a symmetric stretching mode q_1 and a symmetric bending mode q_2, which transform as A_1, while an asymmetric stretching mode q_3 transforms as B_2 (symmetry aspects of ro-vibrational calculations are considered in Chapter 6).

The main advantage of the normal mode is their generality. Not only the construction of the $\boldsymbol{l}$-matrix but also the construction of the KEO can be efficiently generalised as a black-box algorithm applicable for arbitrary systems. This property is especially attractive for implementations as numerical algorithms. The symmetry adaptation of the coordinates, properties and basis functions can be automatised as well, which is another important advantage of the normal-mode methodology. Among the disadvantages of the normal coordinates as rectilinear coordinates (including the linearised coordinates) is their poor (or even lack of) performance for non-rigid, large-amplitude motions. Indeed, the definition of the normal modes is based on an expansion around a single equilibrium structure and therefore is not (directly) suitable for systems with multiple minima. They can (and often do) leave the true domain of the problem. Normal modes are also not as physically intuitive as, e.g., valence (or valence-like linearised) coordinates. Large-amplitude motions, vibrations with well-separated local mode characters and multiple minima are usually better represented by geometrically defined coordinates.

The normal-mode structural matrix $\boldsymbol{l}$ does not usually allow for analytic solution, only perhaps for some simplest cases of (linear) diatomics or (planar) triatomics, not least because of the ortho-normality condition in Eq. (2.72), which normally leads to systems of cumbersome transcendental equations, but also because of the condition for a fully uncoupled quadratic potential $V(\boldsymbol{q})$ in Eq. (2.73), which is system dependent.

The (numerical) construction of the normal-mode representation can be integrated into the linearised coordinates as follows. Once the non-orthogonal linearised matrix $\boldsymbol{A}$ is generated as part of the linearised coordinates protocol (e.g. as an inverse of $\boldsymbol{B}$ in Eqs. (2.64, 2.65)), it can be (numerically) orthogonalised by diagonalising the overlap matrix $\boldsymbol{S}$:

$$S_{\lambda,\lambda'} = \sum_{n,\alpha} A_{n,\alpha,\lambda} A_{n,\alpha,\lambda'}\, m_n, \tag{2.75}$$

where the factor m_n is introduced to account for the difference in the definitions of $\boldsymbol{A}$ and $\boldsymbol{l}$ in Eqs. (2.51) and (2.67), respectively. The orthogonal structural matrix $\tilde{\boldsymbol{l}}$ can be found as a unitary transformation by diagonalising $\boldsymbol{S}$ in Eq. (2.75) to form the rectilinear coordinates $\tilde{\boldsymbol{q}}$ as given by

$$\tilde{q}_\lambda = \sum_{n,\alpha} \sqrt{m_n}\, \tilde{l}_{n\alpha,\lambda} \Delta r_{n\alpha}.$$

The last step is to satisfy the normal-mode condition 4 for the uncoupled harmonic oscillators in Eq. (2.73), which can be achieved using a similar numerical method by (numerically) diagonalising the quadratic force constants matrix $\boldsymbol{F}$:

$$\tilde{F}_{\lambda,\lambda'} = \left(\frac{\partial^2 V(\tilde{\boldsymbol{q}})}{\partial \tilde{q}_\lambda \partial \tilde{q}_{\lambda'}} \right)$$

in the representation of $\tilde{\boldsymbol{q}}$ to obtain the final orthogonal representation for the mass-dependent normal-mode coordinates $\boldsymbol{q}$ as in Eq. (2.74).

In electronic structure calculations, normal modes are constructed as eigenvectors of the so-called Hessian matrix, defined as mass-weighted Cartesian gradients of the potential energy for each atom as given by

$$\mathcal{H}_{i,j} = \frac{1}{\sqrt{m_i}\sqrt{m_j}} \frac{\partial^2 V}{\partial x_i \partial x_j},$$

where x_i and x_j are the generalised $3N$ Cartesian coordinates of the nuclei $\{R_{1,X}, R_{1,Y}, R_{1,Z}, \ldots, R_{N,X}, R_{N,Y}, R_{N,Z}, \ldots\}$ in the lab-frame. This approach guarantees the orthogonality of coordinates, which together with the diagonal form of the potential energy function are the two central conditions of the normal-mode coordinates. The gradients are commonly evaluated using the finite differences, which are subject to the centre-of-mass and Eckart conditions satisfied for each finite coordinate displacement.

2.5.2 Vibrational coordinates for linear molecules

Linear molecules are always a special case requiring, at least in principle, two rotational and $3N-5$ vibrational degrees of freedom. Strictly speaking, only a diatomic molecule is truly linear at any arbitrary geometry. Polyatomic linear molecules might be considered linear at equilibrium, but an arbitrary instantaneous nuclear configuration of a vibrating molecule does not have to be linear and in principle requires a full set of $3N-6$ coordinates to characterise their relative positions and three Euler angles to specify their orientation in space. However, any attempts to introduce an azimuthal Euler angle ϕ would fail at linearity, thus breaking the coordinate transformation in Eq. (2.4). That is why a description of this degree of freedom has to be a part of the $(3N-5)$ vibrational set. In this book, we call molecules 'linear' if they have linear geometry at equilibrium, while molecules that can reach a linear configuration of nuclei as part of the vibrational motion are called 'quasi-linear'. More generally, any rotation of two bonds (including internal rotation) is undefined if they form a linear configuration.

The rectilinear $3N-5$ vibrational coordinates (linearised or normal mode) for a linear molecule can be constructed using the same methodology as for the $3N-6$ case, with the following modifications. Assuming that the equilibrium configuration is chosen along the z axis and hence $r^{\rm e}_{nx} = r^{\rm e}_{ny} = 0$, there are only two non-trivial Eckart conditions in Eq. (2.71), for x and y, as follows:

$$\sum_n \sqrt{m_n} r^{\rm e}_{nz} l_{ny,\lambda} = 0, \qquad (2.76)$$

$$\sum_n \sqrt{m_n} r^{\rm e}_{nz} l_{nx,\lambda} = 0. \qquad (2.77)$$

For the normal modes, the ortho-normality condition is the same as in Eq. (2.72), so is the condition of the uncoupled harmonic potential function in Eq. (2.73), just with the number of degrees of freedom ranging from 1 to $M = 3N-5$. Thus, at least formally, no modifications are required for the normal-mode coordinates for the $3N-5$ case of a linear molecule.

The definition of the non-rectilinear coordinates is more subtle. Consider, for example, the linear molecule C_2H_2 shown in Fig. 2.16. The three stretching vibrational coordinates are, e.g., the bond lengths C–C, C–H_1 and C–H_2 (or their displacements from equilibrium). The relative orientation of the CH_1 and CH_2 bonds for the given deformation angles α_1 and α_2 can be characterised by the rotational angles γ_1 and γ_2 about the C–C bond. Here, $\gamma_1+\gamma_2$ is the dihedral angle δ between the two planes H_1-C-C and C-C-H_2, while $\phi = \gamma_1 - \gamma_2$ provides the orientation of the axis x around the molecular axis z. The angles γ_1 and γ_2 are undefined when α_1 and α_2 reach 180°, which leads to singularities in the KEO.

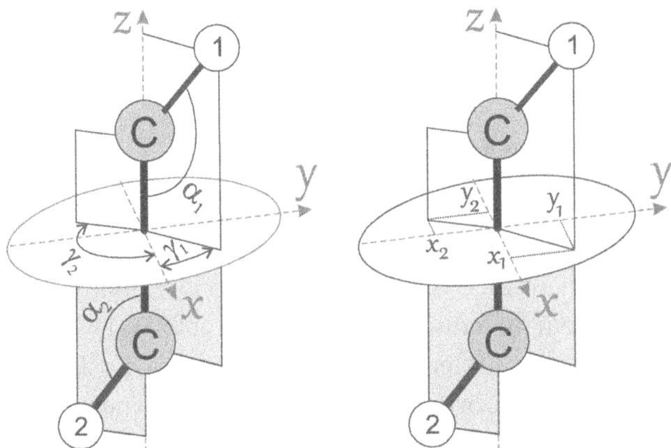

Figure 2.16 Two representations of the angular motion in C_2H_2: using the angles α_1, α_2, γ_1 and γ_2 (left) and four degenerate linearised coordinates Δx_1, Δy_1, Δx_2 and Δy_2 (right).

An alternative representation of the orientations of CH_1 and CH_2 which are fully defined at any configurations is via four Cartesian displacements Δx_1, Δy_1, Δx_2 and Δy_2 from the linear configuration as shown in the right display of Fig. 2.16. In order to remove dependence of the displacements Δx_i and Δy_i on the corresponding bond lengths, it is instructive to normalise them as follows. Consider two bond vectors $\boldsymbol{u}_i$ and $\boldsymbol{v}_i$ (using notation of Fig. 2.8) with an inter-bond angle α_i forming a plane with the unit norm vector $\boldsymbol{n}$ given by

$$\boldsymbol{n} = \frac{\boldsymbol{u} \times \boldsymbol{v}}{|\boldsymbol{u}||\boldsymbol{v}|}.$$

The two (x, y) components of $\boldsymbol{n}$ form a 2D rectilinear representation for the description of the orientation of the bond vector $\boldsymbol{u}_i$ around the bond vector $\boldsymbol{v}_i$ as given by

$$\begin{aligned} \xi_{i,x} &= n_x, \\ \xi_{i,y} &= n_y. \end{aligned}$$

For an N-atomic linear chain molecule with $N-1$ bonds and $N-2$ inter-bond angles (and thus $N-2$ pairs of $\xi^{\ell}_{\lambda_x}, \xi^{\ell}_{\lambda_y}$), this gives exactly $N-1+2(N-2) = 3N-5$ independent vibrational degrees of freedom.

A more detailed illustration of non-singular coordinates used for a linear triatomic molecule CO_2 will be given in Section 4.4.1; see also Fig. 4.7.

Rectilinear coordinates are only suitable around the equilibrium for describing large displacements, while curvilinear coordinates would be more natural, efficient and even correct, e.g. an angle α for XY_2 (Fig. 2.6) or angles α_1, α_2 and δ_{1234} for HOOH in Fig. 2.7, despite the ensuing singularities. We will show in Section 4.1 how specially designed basis sets can be used to cancel the associated singularities.

2.6 OTHER MOLECULES

2.6.1 Rigid tetratomics of the XY_3-type (phosphine)

Here we consider a rigid pyramidal tetratomic molecule XY_3, characterised by the $\mathcal{C}_{3v}(M)$ molecular symmetry group; see Fig. 2.17 with PH_3 as an example. The natural choice for the internal coordinates $\boldsymbol{\xi} = \{\xi_1, \xi_2, \dots \xi_6\}$ is the valence coordinates r_1, r_2 and r_3 (bond lengths) and α_{12}, α_{13} and α_{23} (inter-bond angles), or their displacements from the equilibrium values.

The Cartesian coordinates of the equilibrium structure of a pyramidal tetratomic molecule PH_3 with the centre at the central atom X (i.e. before the centre-of-mass shift) with the z axis along the axis of symmetry are given by

$$
\begin{aligned}
r_{P,x} &= 0, & r_{H_1,x} &= r_e \sin\rho_e, \\
r_{P,y} &= 0, & r_{H_1,y} &= 0, \\
r_{P,z} &= -\frac{3\,m_H\,r_e\cos\rho_e}{3m_H + m_P}, & r_{H_1,z} = r_{H_2,z} = r_{H_3,z} &= \frac{m_P r_e \cos\rho_e}{3m_H + m_P}, \\
r_{H_2,x} = r_{H_3,x} &= -\frac{r_e\sin\rho_e}{2}, & r_{H_2,y} = -r_{H_3,y} &= \frac{\sqrt{3}r_e\sin\rho_e}{2}, \\
r_{H_2,x} &= -\frac{r_e\sin\rho_e}{2},
\end{aligned}
$$

where ρ_e is the angle between the bond P–H$_i$ and the symmetry axis C_3 for the equilibrium structure

$$
\sin\rho_e = \frac{2}{\sqrt{3}}\sin(\alpha_e/2),
$$

and α_e and r_e are the corresponding equilibrium values of the bond length and inter-bond angles. This structure represents the PAS as the equilibrium frame.

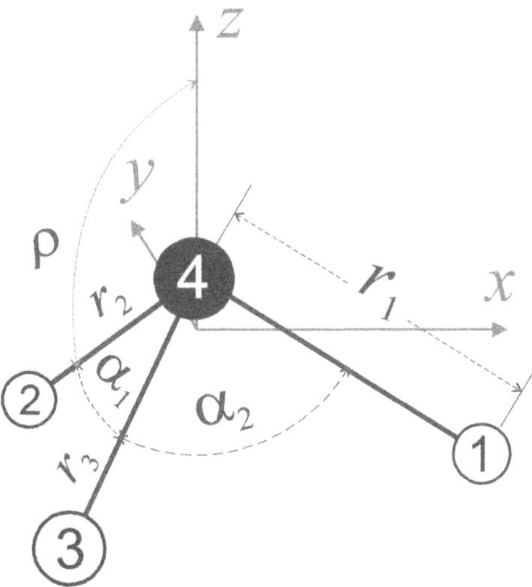

Figure 2.17 The valence coordinates used to define vibrations of a rigid PH_3 molecule.

Linearised coordinates ξ_i^ℓ for PH_3 can now be introduced using rules defined in Section 2.5, i.e. by expanding the definition of the valence coordinates r_i and α_i ($\boldsymbol{\xi}$) from Eqs. (2.52, 2.54) in terms of the Cartesian displacements $\Delta r_{n\alpha}$ in Eq. (2.51) and truncating after 1st order:

$$\xi_\lambda^\ell = \xi_\lambda^{\mathrm{e}} + \sum_{n\alpha} A_{\lambda,n\alpha} \Delta r_{n,\alpha},$$

where $A_{\lambda,n\alpha}$ are the solutions of Eqs. (2.61, 2.62) and $B_{n\alpha,\lambda}$ are obtained as the first derivatives of the valence coordinates with respect to $r_{n\alpha}$ from Eqs. (2.42, 2.43). The elements $A_{\lambda,n\alpha}$ are obtained numerically, and therefore in practice there is no need to have them in their explicit analytic form.

2.6.2 Non-rigid XY_3 (ammonia)

Let us consider the ammonia molecule (NH_3) as an example of a non-rigid XY_3-type molecule. Ammonia also has a pyramidal structure (see Fig. 2.17) characterised by a relatively small barrier to planarity which leads to the important splitting of the inversion level of about 0.78 cm^{-1} in the ground vibrational state. The shape of the potential barrier in the ammonia inversion motion with the height of about 1800 cm^{-1} (shown in Fig. 2.18) is responsible for this tunnelling. The umbrella coordinate plays an important role in the vibrational motion of NH_3 and therefore has to be treated explicitly, which complicates the selection of the vibrational coordinates for this molecule. Even though NH_3 seems to have a similar structure to PH_3, we cannot use the same set of bending coordinates ($\alpha_1, \alpha_2, \alpha_3$) for the reason

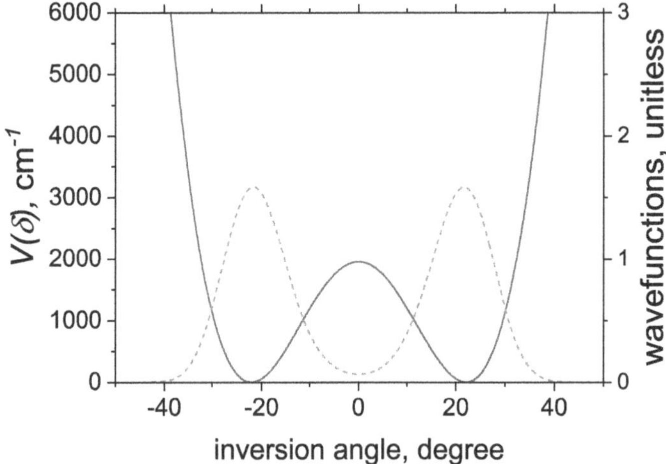

Figure 2.18 NH_3 potential energy function (solid line) along the inversion coordinate δ, with $\delta = 0$ at the planar configuration and the symmetric ground state inversion wavefunction (dashed line).

that these three coordinates do not distinguish the two mirror images of the molecular umbrella configuration relative to the planar structure as shown in Fig. 2.19. Indeed, the values of the inter-bond angles do not depend on the position of the nuclei relative to the planar configuration, and a special, umbrella-type coordinate needs to be introduced as part of the angular description. This is not an issue for PH_3 with a high value of the inversion barrier as long as the tunnelling can be ignored.

There are several ways to define the umbrella coordinate that distinguishes the upper/lower positions of the molecular inversion. For example, when it is in the pyramid configuration (all bond lengths are the same as well as the bond angles), the umbrella motion can be measured as an angle ρ between the z axis, which coincides with the C_3 axis of symmetry and one of the molecular bond (90° at planarity; see Fig. 2.17). Alternatively, it can be defined as an angle between the plane and a molecular bond (δ =0° at planarity); see Fig. 2.19. For the latter, δ is negative when the molecule is below the planar configuration and positive above. The situation is a bit more complicated in a general case of instantaneous configuration when the molecule is not necessarily in an equilateral pyramidal structure. There is no symmetry axis, and the angles between the bonds and the z axis are all different. In this case one can use a trisector $\boldsymbol{n}$ as a reference axis, see Fig. 2.19, defined as a normalised average of the normals to the three H_i-N-H_j planes as follows:

$$\boldsymbol{n} = \frac{\boldsymbol{R}_1 \times \boldsymbol{R}_2 + \boldsymbol{R}_2 \times \boldsymbol{R}_3 + \boldsymbol{R}_3 \times \boldsymbol{R}_1}{|\boldsymbol{R}_1 \times \boldsymbol{R}_2 + \boldsymbol{R}_2 \times \boldsymbol{R}_3 + \boldsymbol{R}_3 \times \boldsymbol{R}_1|}$$

where

$$\boldsymbol{R}_i = \boldsymbol{r}_i - \boldsymbol{r}_4, \quad i = 1, 2, 3.$$

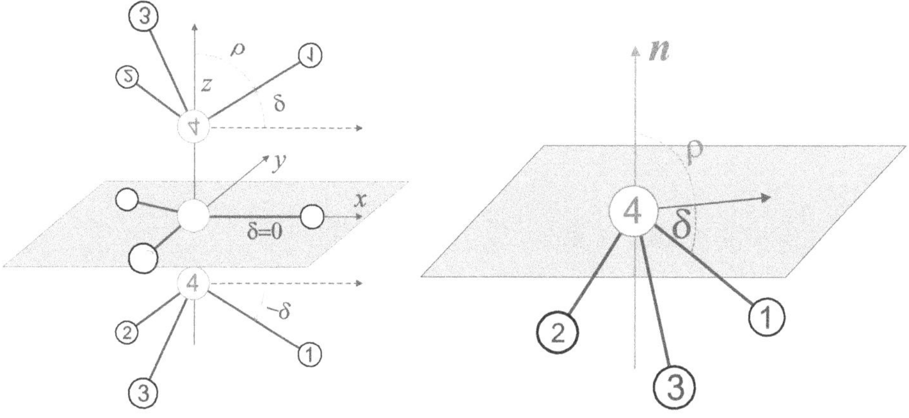

Figure 2.19 Mirror images of the umbrella configuration of NH_3. The inversion angle δ is positive when the three hydrogen atoms are above the molecular plane and negative when they are below the plane. The vector $\boldsymbol{n}$ is the trisector used for the instantaneous configuration of a non-equilateral pyramid.

The umbrella coordinate δ is defined as the projection of (any) $\boldsymbol{R}_i$ on the plane with unit normal vector $\boldsymbol{n}$:

$$\sin\delta = \boldsymbol{n}\cdot\frac{\boldsymbol{R}_i}{R_i}.$$

Using inter-bond angles α_i to define the instantaneous configuration of NH_3, for $\sin\delta$ one can obtain

$$\sin\delta = \pm\frac{a}{b},$$

where

$$a^2 = 1 - \cos\alpha_1^2 - \cos\alpha_2^2 - \cos\alpha_3^2 + 2\cos\alpha_1\cos\alpha_2\cos\alpha_3$$

and

$$b^2 = \sin\alpha_3^2 + \sin\alpha_2^2 + \sin\alpha_1^2 + 2\cos\alpha_3\cos\alpha_1 - 2\cos\alpha_2 + \\ 2\cos\alpha_2\cos\alpha_3 - 2\cos\alpha_1 + 2\cos\alpha_2\cos\alpha_1 - 2\cos\alpha_3.$$

A sensible, symmetrically adapted choice for angular coordinates comprises the standard E-symmetry combinations of the inter-bond angles as

$$S_a = \frac{1}{\sqrt{6}}(2\alpha_1 - \alpha_2 - \alpha_3), \\ S_b = \frac{1}{\sqrt{2}}(\alpha_2 - \alpha_3). \tag{2.78}$$

The complete set of the $3N - 6 = 6$ vibrational coordinates is then given by

$$\xi_1 = \Delta r_1, \\ \xi_2 = \Delta r_2, \\ \xi_3 = \Delta r_3, \tag{2.79}$$

$$\xi_4 = \frac{1}{\sqrt{6}}\left[2\Delta\alpha_1 - \Delta\alpha_2 - \Delta\alpha_3\right], \\ \xi_5 = \frac{1}{\sqrt{2}}\left[\Delta\alpha_2 - \Delta\alpha_3\right], \tag{2.80}$$

$$\xi_6 = \delta. \tag{2.81}$$

The stretching coordinates can also be symmetrised:

$$\xi_1 = \frac{1}{\sqrt{3}}\left[\Delta r_1 + \Delta r_2 + \Delta r_3\right], \\ \xi_2 = \frac{1}{\sqrt{6}}\left[2\Delta r_1 - \Delta r_2 - \Delta r_3\right], \\ \xi_3 = \frac{1}{\sqrt{2}}\left[\Delta r_2 - \Delta r_3\right].$$

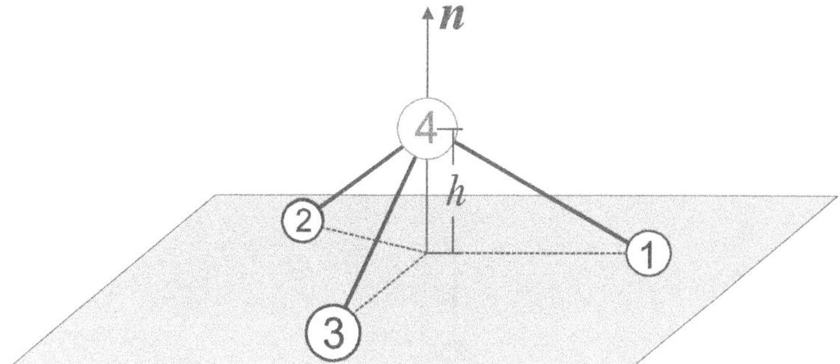

Figure 2.20 The definition of the umbrella coordinate h for NH_3. The vector $\boldsymbol{n}$ is the normal to the plane formed by the three H nuclei.

The non-symmetrised, 'local-mode' option is chemically more intuitive for separating the individual stretching modes in a molecule which can be sometimes useful.

An alternative measure of the umbrella mode is the distance h between the central nucleus N and the plane formed by the three nuclei H (see Fig. 2.20) with $h = 0$ when the molecule has the planar structure given by

$$h = \frac{D}{\sqrt{A^2 + B^2 + C^2}},$$

where the parameters A, B, C and D satisfy the plane condition

$$Ax + By + Cz + D = \begin{vmatrix} x & y & z & 1 \\ \mathrm{NH}_{1x} & \mathrm{NH}_{1y} & \mathrm{NH}_{1z} & 1 \\ \mathrm{NH}_{2x} & \mathrm{NH}_{2y} & \mathrm{NH}_{2z} & 1 \\ \mathrm{NH}_{3x} & \mathrm{NH}_{3y} & \mathrm{NH}_{3z} & 1 \end{vmatrix}$$

and h changes sign when passing through the planar configuration.

2.7 NON-LINEAR RIGID TETRATOMIC MOLECULES: H_2CO TYPE

One of the simplest tetratomic molecules to formulate the KEO and to serve as an illustration is a semi-rigid planar molecule of the ZXY_2 type as for formaldehyde (H_2CO). The natural choice of the vibrational coordinates is the three bond lengths C–H_1 (r_1), C–H_2 (r_2) and C–O (r_3); two bond angles α_1 and α_2; and a dihedral angle θ as shown in Fig. 2.21.

Let us first use these valence, curvilinear coordinates r_1, r_2, r_3, α_1, α_2 and θ to introduce a geometrically defined frame. For a molecule with a backbone such as C–O, it is instructive to place the z axis along this bond (before the CM shift) and the x axis in the plane containing z and bisecting the book angle θ between O–C–H_1 and O–C–H_2; see Fig. 2.21. The y axis is then oriented in the right-handed sense. When defining the xyz frame Cartesian coordinates, it is convenient to initially place a reference coordinate centre at the atom C (defined by the position vectors $\boldsymbol{a}_n$) and then apply the CM shift $\boldsymbol{R}^{(\mathrm{CM})}$. For an arbitrary instantaneous configuration of

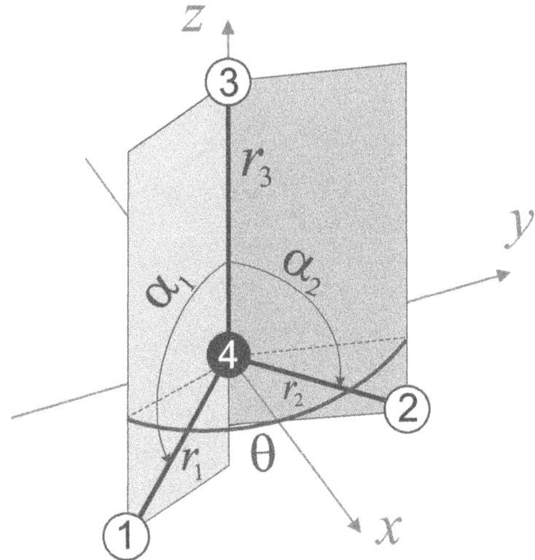

Figure 2.21 The geometric xyz frame and valence coordinates used for the H_2CO molecule. The z axis is along the C–O bond; the x axis is parallel to the bisector of the dihedral angle θ; α_1 and α_2 are the two bending modes; r_1, r_2 and r_3 are the stretching coordinates.

nuclei as defined by r_1, r_2, r_3, α_2, α_3 and θ, the body-frame Cartesian coordinates are given by

$$r_{n\alpha} = a_{n\alpha} + R_\alpha^{(\mathrm{CM})}, \tag{2.82}$$

where

$$a_{4x} = 0, \qquad a_{3y} = 0, \tag{2.83}$$

$$a_{4z} = 0, \qquad a_{3x} = 0, \tag{2.84}$$

$$a_{4y} = 0, \qquad a_{3z} = r_3, \tag{2.85}$$

$$a_{1x} = r_1 \sin\alpha_1 \cos\left(\frac{\theta}{2}\right), \qquad a_{2x} = r_2 \sin\alpha_2 \cos\left(\frac{\theta}{2}\right), \tag{2.86}$$

$$a_{1y} = -r_1 \sin\alpha_1 \sin\left(\frac{\theta}{2}\right), \qquad a_{2y} = -r_2 \sin\alpha_2 \sin\left(\frac{\theta}{2}\right), \tag{2.87}$$

$$a_{1z} = r_1 \cos\alpha_1, \qquad a_{2z} = r_2 \cos\alpha_2. \tag{2.88}$$

The centre-of-mass shift $\boldsymbol{R}^{(\mathrm{CM})}$ is given by

$$R_x^{(\mathrm{CM})} = -(r_1 \sin\alpha_1 + r_2 \sin\alpha_2) \cos\left(\frac{\theta}{2}\right) \frac{m_\mathrm{H}}{M_\mathrm{tot}}, \tag{2.89}$$

$$R_y^{(\mathrm{CM})} = (r_1 \sin\alpha_1 - r_2 \sin\alpha_2) \sin\left(\frac{\theta}{2}\right) \frac{m_\mathrm{H}}{M_\mathrm{tot}}, \tag{2.90}$$

$$R_z^{(\mathrm{CM})} = -(r_1 \cos\alpha_1 + r_2 \cos\alpha_2) \frac{m_\mathrm{H}}{M_\mathrm{tot}} - r_3 \frac{m_\mathrm{O}}{M_\mathrm{tot}} \tag{2.91}$$

such that the elements of $\boldsymbol{r}_n$ satisfy the three CM constraints

$$\sum_n m_n r_{n\alpha} = 0,$$

where $M_{\text{tot}} = m_{\text{C}} + m_{\text{O}} + 2m_{\text{H}}$ is the total mass of H_2CO.

These are polyspherical coordinates, defined using a variation of the Z-matrix scheme (see Section 2.4.2) with the x axis chosen to bisect the dihedral angle between the two equivalent atoms H_1 and H_2 to maintain the symmetry.

Apart from the three centre-of-mass conditions, the three orientational constraints (called C_x, C_y and C_z for future reference) can be formulated in terms of the Cartesian coordinates as follows: the bond O to C has zero x (constraint 1) and y (constraint 2) components, and the x axis splits the book angle between H_1–C–O and H_2–C–O into two halves (constraint 3), or

$$C_x = r_{3x} - r_{4x} = 0, \tag{2.92}$$

$$C_y = r_{3y} - r_{4y} = 0, \tag{2.93}$$

$$C_z = \frac{r_{1y} - r_{4y}}{r_{1x} - r_{4x}} + \frac{r_{2y} - r_{4y}}{r_{2x} - r_{4x}} = 0, \tag{2.94}$$

where we used the following identities for $\tan(\theta/2)$ from Eqs. (2.83–2.88):

$$\frac{r_{1y}}{r_{1x}} = -\tan\left(\frac{\theta}{2}\right), \tag{2.95}$$

$$\frac{r_{2y}}{r_{2x}} = \tan\left(\frac{\theta}{2}\right). \tag{2.96}$$

The remaining six $(3N-6)$ conditions can be associated with the definition of the valence coordinates $\boldsymbol{\xi} = \{\xi_1, \xi_2, \dots \xi_6\} = \{r_1, r_2, r_3, \alpha_1, \alpha_2, \theta\}$ in terms of the Cartesian coordinates via Eqs. (2.23–2.25). A simplified, fraction-less alternative of the constraint C_z is given by

$$C_z = (r_{1y} - r_{4y})(r_{2x} - r_{4x}) + (r_{2y} - r_{4y})(r_{1x} - r_{4x}) = 0$$

obtained by multiplying Eq. (2.94) with $(r_{1x} - r_{4x})$ and $(r_{2x} - r_{4x})$.

2.8 NON-RIGID CHAIN MOLECULE HOOH

Hydrogen peroxide (HOOH or H_2O_2) is an asymmetric prolate rotor molecule. It is one of the simplest molecules with internal (torsional) rotation. The valence coordinates describing six independent vibrational degrees of freedom are (see Fig. 2.7) the O–O bond length $R \equiv r_{23}$, two O–H bond lengths $r_1 \equiv r_{12}$ and $r_2 \equiv r_{34}$, two inter-bond angles H–O–H $\alpha_1 \equiv \alpha_{123}$ and $\alpha_2 \equiv \alpha_{234}$ and a dihedral (book) angle H–O–O–H $\tau \equiv \delta_{1234}$. The torsional potential energy function of HOOH has a double-well structure corresponding to two minima, *cis* and *trans* with barrier heights of about 421 cm^{-1} and 3192 cm^{-1}, which is illustrated in Fig. 2.22. In

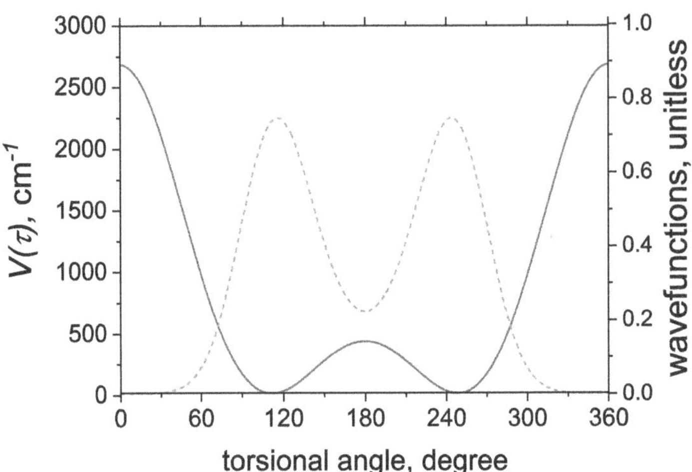

Figure 2.22 Torsional potential energy function of HOOH $V(\tau)$ (solid line) and its ground symmetric state torsional function $\psi_{\text{g.s.}}(\tau)$ (dashed line), where τ is a dihedral angle H–O–O–H between two O–H moieties.

spectroscopic applications it is commonly assumed that HOOH cannot reach the linear configuration, not even for one O–H side. That is, the geometries with $\alpha_1 = 180°$ or $\alpha_2 = 180°$ are considered unfeasible. By spectroscopic convention, HOOH is characterised by six vibrational modes: ν_1 and ν_5 represent the symmetric and asymmetric O-H stretching modes, respectively; ν_3 and ν_6 are the O-H bending modes, respectively; ν_2 represents the O-O stretch and the ν_4 mode represents the torsional excitation.

The main feature of the ro-vibrational treatment of this system is the coupling between the rotational and torsional motions which makes the choice of the body-fixed xyz frame especially important. If we neglect the interaction between the O-H_1 and O-H_2 moieties, they can be effectively described as two free rotors in the LF frame. In our description, this should not be coupled with the overall molecular rotation. In order to uncouple these torsional modes from the molecular rotation, we define a non-rigid reference xyz frame dynamically for any value of τ as follows.

Let us consider the nuclear motion represented as a deviation from a non-rigid reference configuration at each value of the torsional angle τ. The non-rigid xyz frame (see Fig. 2.23) has all structural parameters, the bond lengths for the O-O bond (R), the O-H_1 bond (r_1), the O-H_2 bond (r_2) and the bond angles for O-H_1 (α_1) and O-H_2 (α_2) frozen at their equilibrium values corresponding to the global minimum of the system and the dihedral angle τ as a torsional coordinate varying from 0 to 360°. The x axis is chosen to bisect the dihedral angle τ at any torsional configuration with the z axis pointing up and defined to satisfy the PAS conditions, while the y axis is oriented to form a right-handed system (see Fig. 2.23).

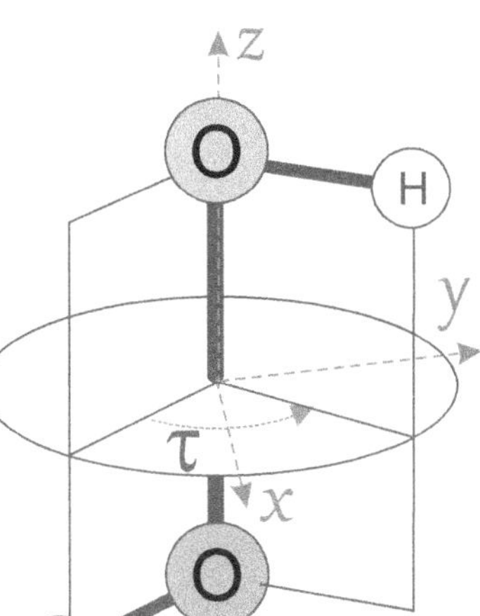

Figure 2.23 PAS for HOOH with the x-axis bisecting the dihedral angle τ.

This dynamic PAS xyz frame with the x axis bisecting the dihedral angle τ has an interesting property. When the two O–H moieties rotate by 180° and the τ angle changes by 360°, the molecule returns to the same nuclear configuration but it is oriented in the opposite direction to its initial position at $\tau = 0°$ relative to the bisector frame xyz. This is illustrated in Fig. 2.24 where we follow the orientations of HOOH relative to the x axis as τ evolves from 0 to 360°. Here we place the two hydrogen atoms at $\tau = 0°$ on top of each other and let them rotate in opposite directions to maintain the x orientation in space unchanged, but always bisecting τ. When τ reaches 360°, the two hydrogen bonds do not return to the original spatial orientation relative to the x axis, pointing in the opposite direction relative to their original orientation. This indicates that our coordinate choice does not fully decouple the torsional and rotational degrees of freedom. As a consequence we must expect that, e.g., the basis set cannot be fully decoupled into a product of the torsional and rotational parts; otherwise they would lead to a phase change of the total wavefunction.

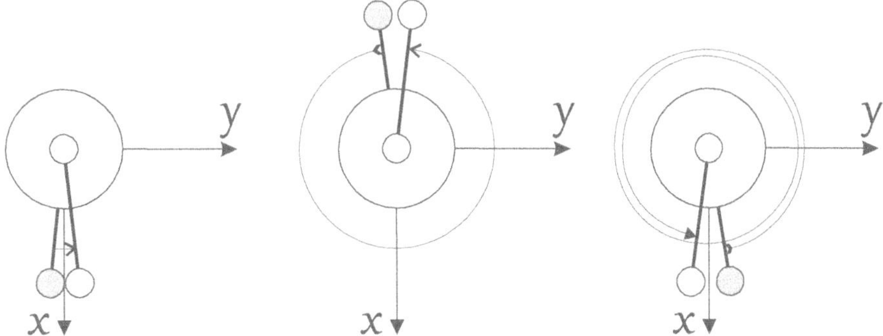

Figure 2.24 PAS with an extended torsional angle $\tau = 0° \ldots 720°$ for HOOH.

There is, however, an elegant solution to this problem. We now extend the torsional angle τ beyond 360° till 720° and thus let the two hydrogens complete their rotation to the original configuration as at $\tau = 0°$ (on top of each other and pointing along x). With this extension it becomes possible to build a rotational-torsional basis set as a product of the rotational (3D rigid rotor) and torsional (defined on $0\ldots720°$) wavefunctions. In this representation $\tau = 0°, 360°$ and $720°$ correspond to the *cis* barrier, while at $\tau = 180°$ and $540°$ the molecule has the *trans* configuration.

The other five degrees of freedom can then be defined as displacements from the equilibrium reference values R_e, r_e and α_e:

$$
\begin{aligned}
\xi_1 &= R - R_e, \\
\xi_2 &= r_1 - r_e, \\
\xi_3 &= r_2 - r_e, \\
\xi_4 &= \alpha_1 - \alpha_e, \\
\xi_5 &= \alpha_2 - \alpha_e,
\end{aligned}
$$

where R_e, r_e and α_e are the corresponding equilibrium values and we also introduce $\xi_6 = \tau$ for consistency.

A more efficient non-rigid reference frame is to use the minimum energy path (MEP). Instead of fixing the non-rigid configuration to the equilibrium values of R_e, r_e and α_e, an MEP frame can be defined in terms of the reaction path as the torsional model with an instantaneous geometry optimised by minimising the potential energy of the molecule along the reaction coordinate τ.

2.9 METHANE AS AN EXAMPLE OF AN XY_4-TYPE TETRAHEDRAL MOLECULE

CH_4 is a tetrahedral five-atomic molecule with nine vibrational degrees of freedom with the $\mathcal{T}_d$ point group symmetry, which consists of five irreducible representations, A_1, A_2, E, F_1 and F_2. A typical molecular axis system for CH_4 at equilibrium is shown in Fig. 2.25 with the Cartesian coordinates given by

$$
\begin{aligned}
H_{1x} &= \frac{r_e}{\sqrt{3}}, & H_{1y} &= -\frac{r_e}{\sqrt{3}}, & H_{1z} &= \frac{r_e}{\sqrt{3}}, \\
H_{2x} &= -\frac{r_e}{\sqrt{3}}, & H_{2y} &= -\frac{r_e}{\sqrt{3}}, & H_{2z} &= -\frac{r_e}{\sqrt{3}}, \\
H_{3x} &= \frac{r_e}{\sqrt{3}}, & H_{3y} &= \frac{r_e}{\sqrt{3}}, & H_{3z} &= -\frac{r_e}{\sqrt{3}}, \\
H_{4x} &= -\frac{r_e}{\sqrt{3}}, & H_{4y} &= -\frac{r_e}{\sqrt{3}}, & H_{4z} &= \frac{r_e}{\sqrt{3}}.
\end{aligned}
$$

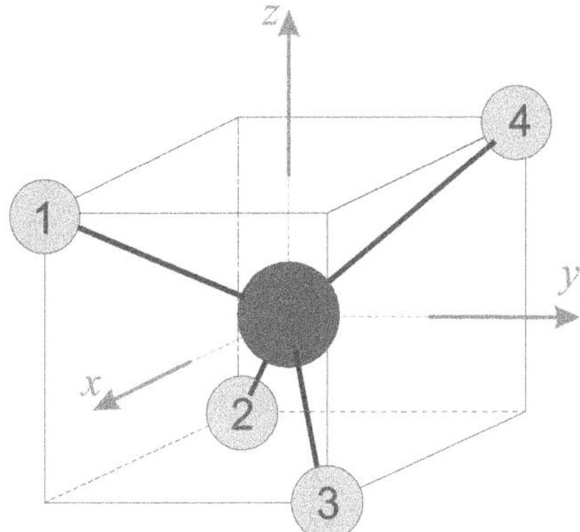

Figure 2.25 Internal coordinate xyz frame of a tetrahedral XY_4 molecule.

The structure of a semi-rigid CH_4 (i.e. ignoring any isomerisation that can occur at higher energies) can be characterised by four bond lengths $r_i \equiv r_{\mathrm{C-H}_i}$ and six inter-bond angles $\alpha_{\mathrm{H}_i-\mathrm{C}-\mathrm{H}_j} = \alpha_{ij}$. For the equilibrium value of the tetrahedral angle α, $\cos(\alpha_{\mathrm{e}}) = -1/\sqrt{3}$ which explains the factor $1/\sqrt{3}$ in the definition of the Cartesian coordinates. There should, however, be only nine independent vibrational degrees of freedom in a five-atomic molecule. One of the inter-bond angles α_{ij} is redundant as there should be only five independent bending vibrations, with the following redundancy condition:

$$\begin{vmatrix} 1 & \cos\alpha_{12} & \cos\alpha_{13} & \cos\alpha_{14} \\ \cos\alpha_{12} & 1 & \cos\alpha_{23} & \cos\alpha_{24} \\ \cos\alpha_{13} & \cos\alpha_{23} & 1 & \cos\alpha_{34} \\ \cos\alpha_{14} & \cos\alpha_{24} & \cos\alpha_{34} & 1 \end{vmatrix} = 0. \tag{2.97}$$

One way to define independent bending modes is to reduce the six inter-bond angles α_{ij} to five symmetry-adapted irreducible combinations, which, together with four bond lengths r_i, form nine independent vibrational modes ξ_i as follows: four stretches

$$\xi_i = r_i, \quad i = 1, 2, 3, 4, \tag{2.98}$$

two E-symmetry bends

$$\begin{aligned} \xi_5^{E_a} &= \frac{1}{\sqrt{12}}(2\alpha_{12} - \alpha_{13} - \alpha_{14} - \alpha_{23} - \alpha_{24} + 2\alpha_{34}), \\ \xi_6^{E_b} &= \frac{1}{2}(\alpha_{13} - \alpha_{14} - \alpha_{23} + \alpha_{24}) \end{aligned} \tag{2.99}$$

and three F-symmetry bends

$$\begin{aligned}
\xi_7^{F_{2x}} &= \frac{1}{\sqrt{2}}(\alpha_{24} - \alpha_{13}), \\
\xi_8^{F_{2y}} &= \frac{1}{\sqrt{2}}(\alpha_{23} - \alpha_{14}), \\
\xi_9^{F_{2z}} &= \frac{1}{\sqrt{2}}(\alpha_{34} - \alpha_{12}),
\end{aligned} \tag{2.100}$$

where the corresponding symmetries of the bending modes are indicated. The stretching modes r_i can also be in principle combined into symmetry-adapted coordinates in $\mathcal{T}_\mathrm{d}$:

$$\begin{aligned}
\xi_1^{A_1} &= \frac{1}{2}(r_1 + r_2 + r_3 + r_4), \\
\xi_2^{F_{2x}} &= \frac{1}{2}(r_1 - r_2 + r_3 - r_4), \\
\xi_3^{F_{2y}} &= \frac{1}{2}(r_1 - r_2 - r_3 + r_4), \\
\xi_4^{F_{2z}} &= \frac{1}{2}(r_1 + r_2 - r_3 - r_4).
\end{aligned} \tag{2.101}$$

MATERIALS USED

Detailed derivation of separation of the translation degrees of freedom in the KEO given in Eq. (2.9) can be found in Bunker and Jensen (1998).

Angle bisector coordinate system was considered by Lukka (1995).

For Jacobi coordinates, see Jepsen and Hirschfelder (1959), Aquilanti and Cavalli (1986) and Wei and Carrington (1997).

The definition and derivatives of the valence coordinates with respect to Cartesian coordinates are from van Schaik et al. (1993); see also Blondel and Karplus (1996).

The constraint for the methane bending coordinates is from Halonen (1997).

The HFCO example is from Leforestier et al. (2001).

For NH_3 umbrella coordinates, see Pesonen et al. (2001), Léonard et al. (2002), Handy et al. (1999) and Špirko (1983).

For the discussion of the inversion barrier and splitting of PH_3, see Sousa-Silva et al. (2016).

EF and derivation of the Eckart conditions are from Sørensen (1979) and Bunker and Jensen (1998) as well as a numerical example of the Eckart frame construction in Section 10.2 of Bunker and Jensen (1998). See Eckart frame with non-rectilinear coordinates in Wang and Carrington (2013).

HOOH barrier is from Benderskii et al. (2001). Extension of the torsional degree of freedom τ to $0° \ldots 720°$ is described in Bunker and Jensen (1998) and Al-Refaie et al. (2016). This is, however, an artificial extension unavoidably leading to some non-physical states, which need to be identified and removed from the final rovibrational solution, when used to solve the corresponding nuclear motion problem; see Al-Refaie et al. (2016).

Z-matrix coordinates are from Lauvergnat and Nauts (2002).

For applications of MEP in the description of torsional motion of chain tetratomics, see Kuhn et al. (1999) and Ovsyannikov et al. (2008).

OTHER USEFUL MATERIAL NOT COVERED IN THIS BOOK

The impact of the body-frame choices on the magnitude of the Coriolis coupling was studied by Sárka et al. (2020) and Sarka et al. (2021).

Hyperspherical coordinates used to present the vibrational motion of a three body system (Carter and Meyer, 1990; Johnson, 1983).

For polyspherical coordinates, see Chapuisat and Iung (1992).

For quadratures, see Press et al. (2007).

Detailed descriptions of Eckart and Eckart-Sayvetz methods (Dymarsky and Kudin, 2005; Pesonen, 2014; Szalay, 2014, 2015a).

Description of usage of Eckart conditions with curvilinear coordinates is in Carter et al. (2009), Wei and Carrington (1997), Wang and Carrington (2013), Sadri et al. (2014), Yachmenev and Yurchenko (2015) and Lauvergnat et al. (2016); see also a review by Császár et al. (2012). Numerical methods developed for solving the Eckart conditions are in Strauss and Pickett (1970), Louck and Galbraith (1976), Dymarsky and Kudin (2005), Sadri et al. (2014), Yachmenev and Yurchenko (2015) and Pesonen (2014).

CHAPTER 3

Kinetic energy operator: Coordinate transformation

In this chapter we discuss how to build a kinetic energy operator (KEO) in the representation of a rotating frame with the vibrations described by internal coordinates as introduced above. The key angle is a numerical construction of the KEO. Our method of choice is the Sørensen method, but some other popular methodologies are also discussed.

Arguably the most traditional way of constructing a KEO in the rotated coordinate system is the so-called Podolsky's trick (Podolsky, 1928), which involves the quantisation of a classical form of the kinetic energy of a molecule in a moving frame. In this book we explore an alternative method, which starts with a quantum form of the KEO in the LF frame and which is then transformed into a moving, body-fixed frame via the chain rules. The lab to body frame coordinate transformation is as in Eq. (2.4), which we repeat here for convenience:

$$R_{n,A} = R_A^{\rm CM} + \sum_{\alpha=x,y,z} S_{A,\alpha}(\theta,\phi,\chi)\, r_{n,\alpha}^{\rm BF}(\boldsymbol{\xi}), \tag{3.1}$$

where $n = 1 \ldots N$, $A = X, Y, Z$ and the LF frame KEO to be transformed is as given in Eq. (2.2):

$$\hat{T} = -\frac{\hbar^2}{2} \sum_{n=1}^{N} \frac{1}{m_n} \left[\partial^2/\partial R_{nX}^2 + \partial^2/\partial R_{nY}^2 + \partial^2/\partial R_{nZ}^2\right]. \tag{3.2}$$

The goal of this chapter is to develop a computational procedure to construct a KEO in the chosen representation of the coordinates and conjugate momenta of the body-fixed frame.

3.1 MOLECULAR KEO IN THE MOVING FRAME

We start with the quantum KEO in the Cartesian LF frame form of Eq. (3.2) (see Fig. 2.1) and apply the coordinate transformation in Eq. (3.1) with the aim of

DOI: 10.1201/9780429154348-3

expressing the nuclear motion Hamiltonian operator in Eq. (2.3) in terms of the $3N$ generalised coordinates

$$\Xi = (R_X^{\rm CM}, R_Y^{\rm CM}, R_Z^{\rm CM}, \theta, \phi, \chi, \xi_1, \xi_2, \ldots, \xi_{3N-6}). \tag{3.3}$$

Let us also introduce the corresponding generalised conjugate momenta $\hat{\Pi}$:

$$\hat{\boldsymbol{\Pi}} = (\hat{P}_X^{\rm CM}, \hat{P}_Y^{\rm CM}, \hat{P}_Z^{\rm CM}, \hat{J}_x, \hat{J}_y, \hat{J}_z, \hat{p}_1, \hat{p}_2, \ldots \hat{p}_{3N-6}). \tag{3.4}$$

In these equations, $R_A^{\rm CM} = \sum_{i=1}^N m_i R_{iA} / \sum_{j=1}^N m_j$ is the A coordinate of the nuclear centre-of-mass ($A = X, Y, Z$); these three coordinates describe the translational motion. The Euler angles (θ, ϕ and χ) define the orientation of the xyz axis system relative to the XYZ system and describe the overall rotation. The vibrational motion is described by the internal coordinates $\xi_1, \xi_2, \ldots, \xi_{3N-6}$, which we represent by the vector $\boldsymbol{\xi}$. $\hat{P}_A^{\rm CM}$ ($A = X, Y, Z$) is the momentum conjugate to the translational coordinate $R_A^{\rm CM}$, ($\hat{J}_x, \hat{J}_y, \hat{J}_z$) are the x, y, z components of the total angular momentum, and $\hat{p}_\lambda = -i\hbar\partial/\partial\xi_\lambda$ ($\lambda = 1, \ldots, 3N-6$) is the momentum conjugate to the vibrational coordinate ξ_λ.

After applying these transformations, as will be shown below, the KEO will have the following general quadratic form in terms of the generalised momenta $\hat{\Pi}_\alpha$:

$$\begin{aligned}
\hat{T} &= \frac{1}{2} \sum_{F=X,Y,Z} \hat{P}_F^{\rm CM} G_{FF} \hat{P}_F^{\rm CM} \\
&+ \frac{1}{2} \sum_{\alpha=x,y,z} \sum_{\alpha'=x,y,z} \hat{J}_\alpha G_{\alpha,\alpha'}(\boldsymbol{\xi}) \hat{J}_{\alpha'} \\
&+ \sum_{\alpha=x,y,z} \sum_{n=1}^{3N-6} \left[\hat{J}_\alpha G_{\alpha,\lambda}(\boldsymbol{\xi}) \hat{p}_\lambda + \hat{p}_\lambda G_{\alpha,\lambda}(\boldsymbol{\xi}) \hat{J}_\alpha\right] \\
&+ \sum_{\lambda=1}^{M} \sum_{\lambda'=1}^{M} \hat{p}_\lambda G_{\lambda,\lambda'}(\boldsymbol{\xi}) \hat{p}_{\lambda'} + U(\boldsymbol{\xi}),
\end{aligned} \tag{3.5}$$

where M is the number of internal (vibrational) degrees of freedom, $3N-6$ or $3N-5$. The kinetic energy factors (i.e. elements of the kinetic matrix $\boldsymbol{G}$) $G_{\alpha,\alpha'}(\boldsymbol{\xi})$, $G_{\alpha,\lambda}(\boldsymbol{\xi})$ and $G_{\lambda,\lambda'}(\boldsymbol{\xi})$ and the pseudo-potential term $U(\boldsymbol{\xi})$ depend on the vibrational coordinates, ξ_λ. We use the notation '$(\boldsymbol{\xi})$' to signify that a function depends on the complete set of vibrational coordinates $\xi_1, \xi_2, \ldots, \xi_M$. The first term in Eq. (3.5) containing $G_{XX} = G_{YY} = G_{ZZ} = 1/\sum_{j=1}^N m_j$ is associated with the translational motion; it is exactly separable from the remaining KEO terms describing the rotation and vibration motions of the nuclei.

For future reference, let us also rewrite this expression in a more compact form

$$\hat{T} = \frac{1}{2} \sum_{i=1}^{3N} \sum_{i'=1}^{3N} \hat{\Pi}_i G_{i,i'}(\boldsymbol{\xi}) \hat{\Pi}_{i'} + U(\boldsymbol{\xi}) \tag{3.6}$$

using the elements of Ξ and $\hat{\boldsymbol{\Pi}}$ from Eqs. (3.3) and (3.4), respectively. The KEO factors $G_{ii'}$ and U do not depend on the Euler angles. The KEO form in Eq. (3.5) corresponds to the so-called Wilson integration volume (Wilson et al., 1955)

$$dV = \sin\theta \, d\xi_1 d\xi_2 \ldots d\xi_M \, d\phi \, d\theta \, d\chi.$$

Here the momenta operators $\hat{\boldsymbol{\Pi}}$ act as normal, i.e. from left to right.

The rotational angular momenta in Eq. (3.5) satisfy the anomalous commutation relation:

$$\hat{J}_\alpha \hat{J}_\beta - \hat{J}_\beta \hat{J}_\alpha = -i\hbar \epsilon_{\alpha\beta\gamma} \hat{J}_\gamma. \tag{3.7}$$

In the following, we show how the elements $G_{i,i'}(\boldsymbol{\xi})$ of the KEO matrix $\boldsymbol{G}(\boldsymbol{\xi})$ can be constructed.

3.2 KEO IN A BODY FRAME: TRANSFORMATIONS OF COORDINATES AND MOMENTA

3.2.1 The t–s formalism

In order to derive a KEO in the quantum form of Eq. (3.6), our method of choice is to start from the original KEO in Eq. (3.2) and transform the conjugate momenta

$$\hat{\boldsymbol{P}} \rightarrow \hat{\boldsymbol{\Pi}}$$

using the chain rule:

$$\frac{\partial}{\partial R_{nA}} = \sum_{i=1}^{3N} \frac{\partial \Xi_i}{\partial R_{nA}} \frac{\partial}{\partial \Xi_i}, \quad n = 1 \ldots N, \quad A = X, Y, Z$$

or

$$\frac{\partial}{\partial R_{nA}} = \sum_{i=1}^{3N} s_{i,nA} \frac{\partial}{\partial \Xi_i}, \tag{3.8}$$

where R_{nA} and Ξ_i are the $3N$ laboratory-fixed and body-fixed coordinates, respectively, and $s_{i,nA}$ are the transformation elements of the Jacobian $\boldsymbol{s}$-matrix given by

$$s_{i,nA} = \frac{\partial \Xi_i}{\partial R_{nA}}. \tag{3.9}$$

Providing the matrix $\boldsymbol{s}$ is invertible, an inverse transformation matrix $\boldsymbol{t}$ can be defined as

$$\boldsymbol{t} = \boldsymbol{s}^{-1}$$

with the elements given by

$$t_{nA,i} = \frac{\partial R_{nA}}{\partial \Xi_i}.$$

The transformation matrices $\boldsymbol{t}$ and $\boldsymbol{s}$ are related to each other as

$$\sum_{i=1}^{3N} t_{n'A',i} s_{i,nA} = \delta_{n,n'} \delta_{A,A'}, \tag{3.10}$$

where $n, n' = 1 \ldots N$, $A = X, Y, Z$ or alternatively using the opposite chain rule:

$$\sum_{n=1}^{N} \sum_{A=X,Y,Z} s_{i,nA}\, t_{nA,i'} = \delta_{i,i'}. \tag{3.11}$$

The $\boldsymbol{s}$-matrix is in fact the $\boldsymbol{B}$-matrix defined by Wilson et al. (1955).

To ensure that the KEO remains Hermitian, the quantum-mechanical transformations of the momenta (inverse and direct) must be given by

$$\hat{P}_{iA} = \frac{1}{2} \sum_{i} \left(s_{i,nA}\, \hat{p}_i + \hat{p}_i\, s_{i,nA} \right), \tag{3.12}$$

$$\hat{p}_{iA} = \frac{1}{2} \sum_{n,A} \left(t_{nA,i}\, \hat{P}_{nA} + \hat{P}_{nA}\, t_{nA,i} \right). \tag{3.13}$$

Now substituting the momenta $\hat{P}_{nA}$ from Eq. (3.12) into the LF frame KEO of Eq. (3.2), after some tedious but straightforward manipulations, the body-frame KEO in Eq. (3.6) can be expressed in terms of the elements of the transformational matrix $\boldsymbol{s}$ as follows. The KEO $G(\boldsymbol{\xi})$-factors are given by the simple scalar-like products of the Cartesian vectors $\boldsymbol{s}_{i,n}$ $(i, j = 1 \ldots 3N)$:

$$G_{i,j} = \sum_{n=1}^{N} \frac{\boldsymbol{s}_{i,n} \boldsymbol{s}_{j,n}}{m_n} = \sum_{n=1}^{N} \sum_{A=X,Y,Z} \frac{s_{i,nA} s_{j,nA}}{m_n}, \tag{3.14}$$

while the pseudo-potential function $U(\boldsymbol{\xi})$ has a more complicated form. It is constructed from the $\boldsymbol{s}_{i,n}$ vectors and their derivatives as given by

$$\begin{aligned} U = & \frac{1}{4} \sum_{n=1}^{N} \sum_{i,j} \frac{1}{m_n} \left(\boldsymbol{s}_{i,n} \left[\hat{p}_i, \left[\hat{p}_j, \boldsymbol{s}_{j,n} \right] \right] + \frac{1}{2} \left[\hat{p}_i, \boldsymbol{s}_{i,n} \right] \left[\hat{p}_j, \boldsymbol{s}_{j,n} \right] \right) \\ = & \frac{1}{4} \sum_{n,A} \sum_{i,j} \frac{1}{m_n} \left(s_{i,nA} \left[\hat{p}_i, \left[\hat{p}_j, s_{j,nA} \right] \right] + \frac{1}{2} \left[\hat{p}_i, s_{i,nA} \right] \left[\hat{p}_j, s_{j,nA} \right] \right), \end{aligned} \tag{3.15}$$

where $n = 1, \ldots, N$ and $A = X, Y, Z$. Here the commutators $[\hat{p}_i, s_{j,nA}]$ have the following properties

$$[\hat{P}_A, \boldsymbol{s}_{A,n}] = 0, \tag{3.16}$$

$$[\hat{J}_\alpha, \boldsymbol{s}_{\alpha',n}] = -i\hbar\, \boldsymbol{e}_\alpha \times \boldsymbol{s}^{\mathrm{rot}}_{\alpha',n}, \tag{3.17}$$

$$[\hat{p}_\lambda, \boldsymbol{s}_{\lambda',n}] = -i\hbar \frac{\partial \boldsymbol{s}^{\mathrm{vib}}_{\lambda',n}}{\partial \xi_\lambda}, \tag{3.18}$$

where $\boldsymbol{e}_\alpha$ is a body-fixed axis vector and we introduce the superscripts 'rot' and 'vib' to indicate the rotational and vibrational components of $\boldsymbol{s}_{i,n}$, $i = x, y, z$ and

$i = 1, 2, 3, \ldots M$, respectively. Using the definition of $\hat{\boldsymbol{p}}$ from Eq. (3.4), Eq. (3.15) can be also expressed in terms of the derivatives $\frac{\partial}{\partial \xi_\lambda}$ as follows:

$$\begin{aligned} U &= \frac{\hbar^2}{4} \sum_i \frac{1}{m_i} \left[\frac{1}{2} \sum_{\alpha\beta\gamma\beta'\gamma'} \epsilon_{\alpha\beta\gamma} \epsilon_{\alpha\beta'\gamma'} s^{\rm rot}_{\beta,i\gamma} s^{\rm rot}_{\beta',i\gamma'} + \sum_{\alpha\beta\gamma} \sum_{\lambda} \epsilon_{\alpha\beta\gamma} s^{\rm vib}_{\lambda,i\gamma} \frac{\partial s^{\rm rot}_{\beta,i\alpha}}{\partial \xi_\lambda} \right. \\ &- \left. \sum_{\alpha} \sum_{\lambda\lambda'} s^{\rm vib}_{\lambda,i\alpha} \frac{\partial^2 s^{\rm vib}_{\lambda',i\alpha}}{\partial \xi_\lambda \xi_{\lambda'}} - \frac{1}{2} \sum_{\alpha} \sum_{\lambda\lambda'} \frac{1}{m_i} \frac{\partial s^{\rm vib}_{\lambda,i\alpha}}{\partial \xi_\lambda} \frac{\partial s^{\rm vib}_{\lambda',i\alpha}}{\partial \xi_{\lambda'}} \right], \end{aligned} \tag{3.19}$$

where $\alpha, \beta, \gamma, \beta'$ and γ' assume values x, y or z and $\lambda, \lambda' = 1, 2, 3, \ldots, M$.

It can be shown that the pseudo-potential term can be expressed via elements of the $\boldsymbol{G}$-matrix as follows:

$$U = \frac{\hbar^2}{32} \sum_{i,j}^{M} \left[\frac{G_{i,j}}{\tilde{G}^2} \frac{\partial \tilde{G}}{\partial \Xi_i} \frac{\partial \tilde{G}}{\partial \Xi_j} - 4 \frac{\partial}{\partial \Xi_i} \left(\frac{G_{i,j}}{\tilde{G}} \frac{\partial \tilde{G}}{\partial \Xi_j} \right) \right], \tag{3.20}$$

where Ξ_i are the generalised coordinates as in Eq. (3.3) and $\tilde{G}$ is given by

$$\tilde{G} = \det(\boldsymbol{G}).$$

Provided that the elements of the transformation matrix $\boldsymbol{s}(\boldsymbol{\xi})$ are known at any instantaneous geometry described by the vibrational coordinates $\boldsymbol{\xi} = \{\xi_1, \ldots, \xi_M\}$, either analytically or numerically, Eqs. (3.14, 3.15) fully define the corresponding KEO. The formulation of the matrix $\boldsymbol{s}(\boldsymbol{\xi})$ is given below.

The translation vectors $\boldsymbol{s}^{(\rm tr)}_{A,n}$ $(A = X, Y, Z)$ are orthogonal to the rotational $\boldsymbol{s}^{(\rm rot)}_{g,n}$ $(g = x, z, y)$ and vibrational $\boldsymbol{s}^{(\rm vib)}_{\lambda,n}$ counterparts. In principle, the vibrational vector can be directly evaluated as derivatives of the internal coordinates with respect to the Cartesian coordinates

$$\boldsymbol{s}^{(\rm vib)}_{i,n} = \frac{\partial \xi_\lambda}{\partial \boldsymbol{R}_n}, \tag{3.21}$$

provided that the functional form $\xi_\lambda = \xi_\lambda(\boldsymbol{R}_n)$ is known. The construction of the rotational vector $\boldsymbol{s}^{(\rm rot)}_{g,n}$ is less straightforward and is discussed below.

It is useful to note that the pure vibrational part of the KEO factors $\mathbf{G}$ does not depend on the choice of the embedding (orientation of the body-fixed axes). This can be deduced from the dot-product form of Eq. (3.14), which is invariant to a unitary transformation $\mathbf{T}$. It can be shown that the pseudo-potential function of vibrational coordinates U also does not depend on the embedding, despite the rotational $\boldsymbol{s}$-matrix $(\boldsymbol{s}_{g,n})$ included in Eq. (3.15). This is consistent with an intuitive observation that the choice of molecular axes should not affect the pure vibrational solution.

Let us follow the formulations by Sørensen (1979) and Watson (2004) and derive the transformation matrix $\boldsymbol{s}$ as an inverse of $\boldsymbol{t}$, i.e. as a solution of Eq. (3.10) or Eq. (3.11).

The elements of the $\boldsymbol{t}$-matrix are obtained by differentiating the coordinate transformation in Eq. (3.1) with respect to the elements of the generalised vector Ξ and are given by

$$\boldsymbol{t}_{n,A} = \boldsymbol{e}_A, \qquad \text{(translation)} \tag{3.22}$$

$$\boldsymbol{t}_{n,\alpha} = \boldsymbol{e}_\alpha \times \boldsymbol{r}_n, \qquad \text{(rotation)} \tag{3.23}$$

$$\boldsymbol{t}_{n,\lambda} = \frac{\partial \boldsymbol{r}_n}{\partial \xi_\lambda}, \qquad \text{(vibration)} \tag{3.24}$$

where $\boldsymbol{e}_A$ and $\boldsymbol{e}_\alpha$ are the corresponding lab-fixed and body-fixed frame axes, respectively.

By projecting the $\mathbf{t}_{n,A}$ vectors on the xyz-frame axes, we obtain

$$\begin{array}{lcll} t_{i\alpha,A} & = & S_{A,\alpha}(\theta,\phi,\chi), & A = X,Y,Z \\ t_{i\alpha,\beta} & = & \sum_\gamma \epsilon_{\alpha\beta\gamma} r_{i\gamma}, & \beta = x,y,z \\ t_{i\alpha,n} & = & \partial r_{i\alpha}/\partial \xi_n, & n = 1,\ldots,M, \end{array} \tag{3.25}$$

where the indices α, β and γ assume values x, y or z; $\epsilon_{\alpha\beta\gamma}$ is an element of the fully antisymmetric tensor; $n = 1, 2, \ldots, 3N-6$; and $r_{i\alpha}$ is a $\boldsymbol{\xi}$-dependent α-component of the position vector of nucleus i in the xyz-axis system. In Eq. (3.25), $S_{A,\alpha}$ are elements of the directional cosine matrix $\boldsymbol{S}(\theta,\phi,\chi)$ given in Eq. (3.1). In ro-vibrational applications, only the rotational and vibrational components of the $\boldsymbol{s}$-matrix are used. Moreover, the translational part of $\boldsymbol{t}$ in Eq. (3.25) is fully decoupled from the rotational and vibrational parts. Therefore, for practical purposes and without any loss of generality, the direction cosines matrix $\boldsymbol{S}(\theta,\phi,\chi)$ in the first line of Eq. (3.25) can be replaced with $\boldsymbol{S}(0,0,0) = \mathbb{1}$, which gives the final form of the transformational matrix suitable for numerical (ro-vibrational) applications:

$$\begin{array}{lcll} t_{i\alpha,\alpha'} & = & \delta_{\alpha',\alpha}, & \alpha' = x,y,z \\ t_{i\alpha,\beta} & = & \sum_\gamma \epsilon_{\alpha\beta\gamma} r_{i\gamma}, & \beta = x,y,z \\ t_{i\alpha,n} & = & \partial r_{i\alpha}/\partial \xi_n, & n = 1,\ldots,M. \end{array} \tag{3.26}$$

This is a general formulation. In order to make it specific for a particular coordinate choice, the coordinate vectors $\boldsymbol{r}_i(\xi)$ need to be defined as a function of the internal coordinates. This includes (i) three conditions for choosing the coordinate centre (always at the centre-of-mass of the molecule, with or without electrons), (ii) three conditions for choosing the body-fixed axes (e.g. Eckart system) and (iii) $3N-6$ conditions for choosing the internal, vibrational coordinates (e.g. normal, linearised or valence). These conditions define the $3N \times 3N$ elements of the $\boldsymbol{t}$-matrix in Eq. (3.26), which then needs to be inverted, preferably numerically, to obtain the $\boldsymbol{s}$-matrix and then the kinetic energy factors $\boldsymbol{G}$ and U according to Eqs. (3.14, 3.15).

3.2.2 An alternative, g-matrix method to derive KEO

Using the properties of the transformational $\boldsymbol{t}$-matrix, the construction of the body frame kinetic matrix $G_{i,j}$ can be formulated in terms of the kinetic matrix $\boldsymbol{g}$ defined as

$$g_{i,i'} = \sum_n m_n \boldsymbol{t}_{n,i} \boldsymbol{t}_{n,i'} = \sum_{nA} m_n t_{nA,i} t_{nA,i'}. \tag{3.27}$$

This $3N \times 3N$ matrix $\boldsymbol{g}$ is the kinetic energy factor that defines the Lagrangian of the system containing N nuclei:

$$L = \frac{1}{2} \sum_{i=1}^{3N} \sum_{i'=1}^{3N} g_{i,i'} \dot{\Xi}_i \dot{\Xi}_{i'} - V(\boldsymbol{\xi}),$$

where Ξ are the generalised coordinates as in Eq. (3.3). The kinetic matrix $\boldsymbol{G}$ is obtained as an inverse of the kinetic $\boldsymbol{g}$

$$\boldsymbol{G} = \boldsymbol{g}^{-1}$$

with U defined via Eq. (3.20).

In the following, we show how $\boldsymbol{G}$ and U can be derived using different methods appropriate for numerical applications, where we consider several popular methodologies, frames and associated coordinates for a robust construction of the KEO. We start with the normal-mode KEO, as this is arguably the most popular coordinate choice.

3.3 NORMAL MODES AND WATSON HAMILTONIAN

J.K.G. Watson (1970) presented a general ro-vibrational Hamiltonian based on the normal modes for non-linear $(3N - 6)$ and linear $(3N - 5)$ molecules. This form is often referred to as a Watson Hamiltonian. A big advantage of the normal modes is that they are based on a set of general rules applicable to (at least in principle) any arbitrary semi-rigid molecule and suitable for numerical or indeed algebraic computer algorithms.

Normal-mode coordinates should be defined around the molecular equilibrium, with the coordinate system placed at the centre-of-mass and (commonly) oriented according to the PAS conditions, i.e. with the equilibrium body-frame coordinates $\boldsymbol{r}_n^{\rm e}$ satisfying Eqs. (2.68, 2.69). The structural, normal-mode parameters $\boldsymbol{l}_{n\alpha}$ (see Section 2.5.1) are constructed using the centre-of-mass, Eckart and orthogonality conditions; see Eqs. (2.70–2.73).

According to Watson (1968), a rotation-vibration KEO in the normal-mode representation for a non-linear polyatomic molecule is given by

$$\hat{T} = \frac{1}{2} \sum_{\alpha,\beta} \left(\hat{J}_\alpha - \hat{\pi}_\alpha \right) \mu_{\alpha,\beta} \left(\hat{J}_\beta - \hat{\pi}_\beta \right) + \frac{1}{2} \sum_\lambda p_\lambda^2 + U(\boldsymbol{q}), \tag{3.28}$$

where $\hat{\pi}_\alpha$ are components of the vibrational angular momentum defined as

$$\hat{\pi}_\alpha = \sum_{\lambda,\lambda'} \zeta^\alpha_{\lambda,\lambda'} q_\lambda \hat{p}_{\lambda'}; \tag{3.29}$$

$\hat{p}_\lambda$ is the vibrational momentum conjugate to the normal coordinate q_λ; $U(\boldsymbol{q})$ is the pseudo-potential function given by

$$U(\boldsymbol{q}) = -\frac{\hbar^2}{8} \sum_\alpha \mu_{\alpha,\alpha} \tag{3.30}$$

and $\zeta^\alpha_{\lambda,\lambda'}$ are the Coriolis coefficients defined as

$$\zeta^\alpha_{\lambda,\lambda'} = -\zeta^\alpha_{\lambda',\lambda} = \sum_{\beta,\gamma} \epsilon_{\alpha\beta\gamma} \sum_n l_{n\beta,\lambda} l_{n\gamma,\lambda'}. \tag{3.31}$$

Here $\alpha, \beta, \gamma = x, y, z$ and $\lambda, \lambda' = 1 \ldots 3N - 6$.

The rotational tensor $\boldsymbol{\mu}$ with elements $\mu_{\alpha,\beta}$ in Eqs. (3.28, 3.30) is the inverse of the tensor $\boldsymbol{J}$,

$$\mu_{\alpha,\alpha'} = \left(J^{-1}\right)_{\alpha,\alpha'}, \tag{3.32}$$

with the tensor elements given by

$$J_{\alpha,\alpha'} = I_{\alpha,\alpha'} - \sum_{\lambda,\lambda',\mu} \zeta^\alpha_{\lambda,\mu} \zeta^{\alpha'}_{\lambda',\mu} q_\lambda q_{\lambda'} \tag{3.33}$$

where $I_{\alpha,\alpha'}$ are elements of the instantaneous moment of inertia tensor

$$I_{\alpha,\alpha'} = I_{\alpha',\alpha} = \sum_{\beta,\gamma,\delta} \epsilon_{\alpha\beta\delta} \epsilon_{\alpha'\gamma\delta} \sum_n m_n r_{n\beta} r_{n\gamma}.$$

The instantaneous values of the normal coordinates q_λ are related to Cartesian displacements $\Delta r_{n\alpha}$ in the body frame via the linear transformation given in Eq. (2.67).

In general, there is no closed, analytic solution for $(\mathbf{J}^{-1})$ in Eq. (3.32) as an inverse of the quadratic function of q_λ. This is the main difficulty of the Watson normal-mode approach. Watson (1968) suggested using a Taylor expansion for $\boldsymbol{\mu}$ and derived the following useful relation:

$$\boldsymbol{\mu} = \boldsymbol{I}_0^{-1} - \boldsymbol{I}_0^{-1}\boldsymbol{a}\boldsymbol{I}_0^{-1} + \frac{3}{4}\boldsymbol{I}_0^{-1}\boldsymbol{a}\boldsymbol{I}_0^{-1}\boldsymbol{a}\boldsymbol{I}_0^{-1} - \frac{1}{2}\boldsymbol{I}_0^{-1}\boldsymbol{a}\boldsymbol{I}_0^{-1}\boldsymbol{a}\boldsymbol{I}_0^{-1}\boldsymbol{a}\boldsymbol{I}_0^{-1} + \ldots \tag{3.34}$$

where the tensor $\boldsymbol{a}$ is given by

$$a^\lambda_{\alpha,\alpha'} = a^\lambda_{\alpha',\alpha} = 2\sum_{\beta,\gamma,\delta} \epsilon_{\alpha\beta\delta} \epsilon_{\alpha'\gamma\delta} \sum_n \sqrt{m_n} r^{\mathrm{e}}_{n\beta} l_{n\gamma,\lambda}$$

and I_0 is the equilibrium moment of inertia tensor, which is a diagonal matrix if the PAS is used for the equilibrium configuration.

It is common to represent the normal-mode KEO also as a Taylor expansion in terms of the vibrational coordinates q_λ. Such a sum-of-product form is especially useful in conjunction with the basis sets represented in an analogous product form. This facilitates the computation of the associated Hamiltonian matrix elements, especially when solving the Schrödinger equation variationally, but also when using perturbation methods. The obvious choice of the basis functions for normal modes is the harmonic oscillators, in most cases one-dimensional, in some cases 2D or even 3D isotropic harmonic oscillators. Alternatively, the tensor $\boldsymbol{\mu}$ can be evaluated numerically at any instantaneous geometry defined by $\boldsymbol{q}$ using Eq. (3.32). The latter is usually utilised when quadrature-integration or discrete variable representation (DVR) methods are used to evaluate matrix elements of $\boldsymbol{\mu}$, which require knowledge of the integrand on a given grid of geometries (quadrature points).

3.4 SØRENSEN APPROACH FOR THE ROTATIONAL $\boldsymbol{S}$-MATRIX

Let us now consider the linearised coordinates as vibrational degrees of freedom and the associated body-fixed frame defined in Eq. (2.50) and assume the KEO form of Eq. (3.6)

$$\hat{T} = \frac{1}{2} \sum_{\lambda,\mu=1}^{M+3} \hat{p}_\lambda G_{\lambda\mu}(\boldsymbol{\xi}) \hat{p}_\mu + U(\boldsymbol{\xi}), \tag{3.35}$$

where the translation modes are dropped.

We can now use the $\boldsymbol{s}$-matrix methodology presented in Section 3.2 and build the corresponding KEO. The elements $s_{\lambda,i\alpha}$ ($\lambda = 1, \ldots, M+3$) can be obtained via inversion of the matrix $\boldsymbol{t}$ (defined in Eq. (3.26)) and then used to construct the kinetic energy elements $G_{\lambda\mu}(\boldsymbol{\xi})$ and U as in Eqs. (3.14, 3.15).

In this section we show how to simplify the $\boldsymbol{t}$–$\boldsymbol{s}$ methodology using the Sørensen approach by contracting the $3N \times 3N$ inversion problem to the rotational part only. This approach is built around the three rotational conditions defining the orientation of the BF frame relative to the LF frame (see Section 2.2.2). Let us assume that these rotational constraints can be formulated as a set of three equations

$$C^{(g)} = C^{(g)}(r_{1x}, r_{1y}, \ldots, r_{Nz}) = 0, \tag{3.36}$$

which we label as $g = 1, 2, 3$. Examples of constraints include Eq. (2.18) (z axis parallel the C–N bond), Eq. (2.19) (Eckart) or Eq. (2.22) (PAS) as discussed in Section 2.3. These constraints can be differentiated with respect to the vibrational coordinates $\boldsymbol{\xi}$:

$$\frac{\partial C^{(g)}}{\partial \xi_\lambda} = 0, \quad \lambda = 1, 2, \ldots, M, \tag{3.37}$$

which we evaluate using the chain rule

$$0 = \frac{\partial C^{(g)}}{\partial \xi_\lambda} = \sum_{n=1}^{N} \sum_{\alpha=x,y,z} \frac{\partial C^{(g)}}{\partial r_{n\alpha}} \frac{\partial r_{n\alpha}}{\partial \xi_\lambda}. \tag{3.38}$$

Elements of the constraint vector $\boldsymbol{c}_n^g$ are defined as

$$c_{g,n\alpha} = \frac{\partial C^{(g)}}{\partial r_{n\alpha}} \tag{3.39}$$

($g, \alpha = x, y, z$ and $n = 1, 2, \ldots, N$), which, according to Eq. (3.38), in turn satisfy the following $3 \times M$ 'orthogonality' conditions

$$\sum_{n=1}^{N} \boldsymbol{c}_{g,n} \cdot \boldsymbol{t}_{n,\lambda} = 0 \tag{3.40}$$

with $\lambda = 1, \ldots, M$. These conditions have the same form as the orthogonality conditions between $\boldsymbol{t}$ and $\boldsymbol{s}$ given in Eq. (3.11), and therefore the rotational $\boldsymbol{s}_{g,n}$ vectors and $\boldsymbol{c}_{g,n}$ can be related via a 3×3 transformation matrix $\eta_{gg'}$

$$s_{g,i\alpha}^{(\mathrm{rot})} = \sum_{g'} \eta_{gg'} c_{g',i\alpha}. \tag{3.41}$$

The constraint vectors $\boldsymbol{c}_{g,n}$ are not orthogonal to the rotational $\boldsymbol{t}_{n,g'}$-vectors ($g' = x, y, z$) and we can define a matrix $\boldsymbol{J}$ with the elements

$$J_{gg'} = \sum_{n=1}^{N} \sum_{\alpha=x,y,z} c_{g,n\alpha} t_{n\alpha,g'}, \tag{3.42}$$

which is an invertible 3×3 matrix and an inverse of $\boldsymbol{\eta}$:

$$\sum_{g''} \eta_{g,g''} J_{g''g'} = \delta_{gg'}, \quad g, g' = x, y, z. \tag{3.43}$$

Hence, the vectors $\boldsymbol{s}_{g,n}^{(\mathrm{rot})}$ are constructed using Eq. (3.41), where $\boldsymbol{\eta}$ is obtained by inverting the 3×3 matrix $\boldsymbol{J}$, which in turn is defined using the constraint vectors $\boldsymbol{c}_i$ and the body-frame vectors $\boldsymbol{t}$, i.e. known quantities. Using this method, which we will refer to as Sørensen's approach, the dimension of the coordinate transformation problem is reduced from $3N$ to 3.

Regarding the vibrational part, $\boldsymbol{s}_{\lambda,n}^{(\mathrm{vib})}$, let us just assume for now that it can be evaluated directly as derivatives of the internal coordinates. Such derivatives are naturally defined for valence (geometrically defined) coordinates introduced in Section 2.4.4 and are also usually accessible for other cases.

Let us now consider examples of constraint vectors for some typical embeddings, starting from the Eckart frame. By differentiating the three Eckart conditions given in Eq. (2.19), the constraint vectors $\boldsymbol{c}_{g,n}$ are obtained as

$$c_{g,n\alpha} = m_n \sum_{\beta=x,y,z} \epsilon_{g\alpha\beta} r_{n\beta}^{\mathrm{e}}, \tag{3.44}$$

where $g = x, y, z$ for $(\alpha, \beta) = (y, z), (z, x)$ and (x, y), respectively; $r_{n\beta}^{\mathrm{e}}$ is the β-Cartesian component (β=x, y, z) of the atom i at the equilibrium configuration and $\epsilon_{g\alpha\beta}$ are elements of the fully antisymmetric Levi-Civita tensor.

The PAS frame is defined by the three conditions (see Eq. (2.22))

$$C^{(g)} = \sum_n m_n r_{n\alpha} r_{n\beta} = 0,$$

where $g = x, y, z$ for $(\alpha, \beta) = (y, z), (z, x)$ and (x, y), respectively. The elements of the PAS constraint $\boldsymbol{c}_n^{(g)}$ vectors are then given by

$$c_{g,n\alpha} = m_n r_{n\beta}.$$

Finally, in the case of a geometrically defined frame of HCN (see Section 2.3.1), the corresponding constraint vectors $\boldsymbol{c}_{g,i}$ are obtained by differentiating Eq. (2.18). The corresponding (non-zero) elements $c_{g,i\alpha}$ are then given by

$$\begin{aligned} c_{1,Nx} &= 1, & c_{1,Cx} &= -1, \\ c_{2,Ny} &= 1, & c_{2,Cy} &= -1, \\ c_{3,Hy} &= 1, & c_{3,Cy} &= -1. \end{aligned}$$

More examples of KEOs derived using Sørensen's methodology will be given below.

3.4.1 Example of derivation of Sørensen vectors for H_2CO

Let us now illustrate the Sørensen approach for an H_2CO molecule as an example using valence coordinates and the bisector molecular frame shown in Fig. 2.21. The three orientational constraints in Eqs. (2.92–2.94) provide the required Sørensen $\boldsymbol{C}$ vectors, which in turn lead to the following non-zero components $\boldsymbol{c}_{g,n}$ obtained as first derivatives of $\boldsymbol{C}$ with respect to $r_{n\alpha}$:

$$c_{x,4,x} = -1, \quad c_{x,3,x} = 1, \tag{3.45}$$

$$c_{y,4,y} = -1, \quad c_{y,3,x} = 1, \tag{3.46}$$

$$c_{z,4,x} = 1, \tag{3.47}$$

$$c_{z,4,x} = \frac{r_{1y} - r_{4y}}{(r_{1x} - r_{4x})^2} + \frac{r_{2y} - r_{4y}}{(r_{2x} - r_{4x})^2}, \tag{3.48}$$

$$c_{z,4,y} = -\frac{1}{r_{1x} - r_{4x}} - \frac{1}{r_{2x} - r_{4x}}, \tag{3.49}$$

$$c_{z,1,x} = -\frac{r_{1y} - r_{4y}}{(r_{1x} - r_{4x})^2}, \quad c_{z,1,y} = \frac{1}{r_{1x} - r_{4x}}, \tag{3.50}$$

$$c_{z,2,x} = -\frac{r_{2y} - r_{4y}}{(r_{2x} - r_{4x})^2}, \quad c_{z,2,y} = \frac{1}{r_{2x} - r_{4x}}. \tag{3.51}$$

Using Eqs. (2.82–2.91), these equations can readily be expressed in terms of r_1, r_2, r_3, α_1, α_2 and θ, for example

$$c_{z,4,x} = -\frac{(-r_1 \sin\alpha_1 + r_2 \sin\alpha_2)\sin\left(\frac{1}{2}\theta\right)}{r_1 r_2 \sin\alpha_1 \sin\alpha_2 \cos^2\left(\frac{1}{2}\theta\right)}.$$

The 3 × 3 matrix $\boldsymbol{J}$ is given by Eq. (3.42) and has the simple form

$$\boldsymbol{J} = \begin{pmatrix} 0 & -r_3 & 0 \\ r_3 & 0 & 0 \\ \frac{1}{\cos\left(\frac{\theta}{2}\right)}(\cot\alpha_1 + \cot\alpha_2) & \frac{\sin\left(\frac{\theta}{2}\right)}{\cos^2\left(\frac{\theta}{2}\right)}(\cot\alpha_1 - \cot\alpha_2) & -\frac{1}{2\cos^2\left(\frac{\theta}{2}\right)} \end{pmatrix} \tag{3.52}$$

which we can invert either analytically or numerically at any given geometry $\boldsymbol{\xi}$ to obtain the matrix $\boldsymbol{\eta}$:

$$\boldsymbol{\eta} = \begin{pmatrix} 0 & \frac{1}{r_1} & 0 \\ -\frac{1}{r_1} & 0 & 0 \\ \sin\left(\frac{\theta}{2}\right)\frac{1}{r_1}(\cot\alpha_1 - \cot\alpha_2) & \cos\left(\frac{\theta}{2}\right)\frac{1}{r_1}(\cot\alpha_1 + \cot\alpha_2) & -\frac{1}{2}\cos^2\left(\frac{\theta}{2}\right) \end{pmatrix}.$$

In turn, the latter gives the rotational $\boldsymbol{s}^{\mathrm{rot}}$ matrix via Eq. (3.41). The vibrational elements of $\boldsymbol{s}^{\mathrm{vib}}$ are then obtained using the general relations for derivatives of the valence coordinates with respect to the Cartesian displacements $\boldsymbol{r}_n$, see Section 2.4.4, where the latter are then expressed using valence coordinates via Eq. (2.82). We do not list these expressions here since they are too lengthy and, most importantly, can be computed numerically, which is our primary motivation. All derivations can be formulated numerically at any given instantaneous configuration of the six modes ξ_i; see Section 3.5.1. We only note the determinant of the matrix $\boldsymbol{t}$ in this case:

$$\det(\boldsymbol{t}) = r_1^2 r_2^2 r_3^2 \sin\alpha_1 \sin\alpha_2.$$

The special points where this determinant becomes zero are $\alpha_1 = 0$ or $\alpha_2 = 0$. Let us also note that the resulting KEO of H_2CO produced using this frame is factorisable into a sum-of-products form, if required.

3.5 NUMERICAL EVALUATION OF KEO AT ANY INSTANTANEOUS GEOMETRY

Already for simple triatomic systems, the final expressions for the KEO factors $\boldsymbol{G}$ can be complicated and this applies even more so for large polyatomic molecules, where analytic derivations even with the help of computer algebra quickly become intractable. The commonly used Eckart conditions are generally not analytically solvable for non-planar/non-linear molecules and thus require a numerical solution. Furthermore, as discussed in Introduction, if the final purpose of the KEOs is to be used in numerical solutions of the Schrödinger equations anyway, there is no special benefit of having the KEO in an analytic form. That is why the modern tendency is to turn to numerically driven approaches, with no analytic derivation of the KEO and to only obtain the KEO numerically. We thus look at our formulation in the light of this motivation.

3.5.1 KEO in geometrically defined coordinates

Let us now discuss how to approach a pure numerical formulation of the KEO when the kinetic factors $\boldsymbol{G}(\boldsymbol{\xi})$ and $U(\boldsymbol{\xi})$ for a given molecular frame (defined by the corresponding three rotational constraints), and for any instantaneous arrangement of nuclei (defined by $\boldsymbol{\xi}$), are constructed and used numerically.

Let us start by considering the example of the non-linear XY_2 molecule in some arbitrary instantaneous configuration of nuclei defined by the internal (vibrational) coordinates $\boldsymbol{\xi} = \{r_1, r_2, \alpha\}$. For this example, we choose the bisector molecular frame xyz; see Fig. 2.1. As was shown in Section 3.2.1, the heart of the $\boldsymbol{t}$–$\boldsymbol{s}$ formalism is the $3N \times 3N$ matrix $\boldsymbol{t}$ defined via Eq. (3.26). The translation components of $\boldsymbol{t}$ given in Eq. (3.26) (rows 1,2,3) are decoupled from rotational and vibrational ones and can be assumed to be $\delta_{\alpha,\alpha'}$ ($\alpha = x, y, z$). The rotational components (rows 4,5,6) are constructed from Cartesian coordinates vectors $\boldsymbol{r}_n$ crossed with the body-fixed coordinate axes $\boldsymbol{e}_g$ ($g = x, y, z$). The vibrational components of $\boldsymbol{t}$ (rows 7,...9) are the derivatives of $\boldsymbol{r}_n$ with respect to internal coordinates $\boldsymbol{\xi} = r_1, r_2, \alpha$.

Once the matrix elements $t_{n\alpha,i}$ are defined, a $3N \times 3N$ matrix $\boldsymbol{s}$ is readily obtained as the inverse of $\boldsymbol{t}$; see Eq. (3.11). Even though inverting $\boldsymbol{t}$ analytically might not be easy or even possible, numerically this is usually not a problem, at least for invertible matrices (exceptions include linear configurations of nuclei leading to singularities in the KEO). There exist many efficient numerical libraries suitable for inverting relatively small ($3N \times 3N$) matrices $\boldsymbol{t}$ required for building the transformation $\boldsymbol{s}$-matrices (and therefore KEOs) numerically. A numerical scheme for evaluating the kinetic energy factors $\boldsymbol{G}$ at a specific instantaneous configuration of the nuclei defined by $\boldsymbol{\xi}$ can be summarised as follows:

$$\begin{array}{lcccccccc} \text{Eqs.} & & & (3.26) & & (3.11) & & (3.14) & \\ \boldsymbol{\xi} & \rightarrow & (\boldsymbol{r}_n(\boldsymbol{\xi}), \frac{\partial \boldsymbol{r}_n}{\partial \xi_i}) & \rightarrow & \boldsymbol{t}_{n,i}(\boldsymbol{\xi}) & \rightarrow & \boldsymbol{s}_{i',n'}(\boldsymbol{\xi}) & \rightarrow & G_{i,i''}(\boldsymbol{\xi}). \end{array}$$

The pseudo-potential function U given in Eq. (3.15) requires the first and second derivatives of $\boldsymbol{s}_{n,\lambda}$ with respect to $\boldsymbol{\xi}$, which can be related to the derivatives of $\boldsymbol{t}$ as follows:

$$\frac{\partial \boldsymbol{s}}{\partial \xi_i} = -\boldsymbol{s}\frac{\partial \boldsymbol{t}}{\partial \xi_i}\boldsymbol{s} \tag{3.53}$$

$$\frac{\partial^2 \boldsymbol{s}}{\partial \xi_i \partial \xi_j} = \boldsymbol{s}\frac{\partial \boldsymbol{t}}{\partial \xi_i}\boldsymbol{s}\frac{\partial \boldsymbol{t}}{\partial \xi_j}\boldsymbol{s} + \boldsymbol{s}\frac{\partial \boldsymbol{t}}{\partial \xi_j}\boldsymbol{s}\frac{\partial \boldsymbol{t}}{\partial \xi_i}\boldsymbol{s} - \boldsymbol{s}\frac{\partial^2 \boldsymbol{t}}{\partial \xi_i \partial \xi_j}\boldsymbol{s}. \tag{3.54}$$

Considering that the vibrational elements $t_{n\alpha,\lambda}$ already contain the first derivative of $\boldsymbol{r}_n$, up to the third derivatives of the $\boldsymbol{r}_n^{(\mathrm{BF})}$ are needed to define U:

$$\boldsymbol{\xi} = \{r_1, r_2, \alpha\} \rightarrow \boldsymbol{r}_n(\boldsymbol{\xi}), \frac{\partial \boldsymbol{r}_n}{\partial \boldsymbol{\xi}}, \frac{\partial^2 \boldsymbol{r}_n}{\partial \boldsymbol{\xi} \partial \boldsymbol{\xi}}, \frac{\partial^3 \boldsymbol{r}_n}{\partial \boldsymbol{\xi} \partial \boldsymbol{\xi} \partial \boldsymbol{\xi}}, \rightarrow \cdots \rightarrow U(\boldsymbol{\xi}).$$

Depending on the type of embedding, the derivatives of the rotational vectors $\boldsymbol{t}_{n,g}$ in Eq. (3.26) with respect to $\boldsymbol{\xi}$ can be evaluated analytically using the definition of the corresponding coordinates.

It is even more straightforward to evaluate derivatives numerically via the finite differences method. For example, applying the central 2-point stencil formula, for the first derivative of $\boldsymbol{s}$ we obtain:

$$\frac{\partial \boldsymbol{s}(\boldsymbol{\xi})}{\partial \xi_i} = \frac{\boldsymbol{s}(\xi_1, \ldots, \xi_i + \Delta, \ldots) - \boldsymbol{s}(\xi_1, \ldots, \xi_i - \Delta, \ldots)}{2\Delta}, \tag{3.55}$$

where Δ is a small displacement. The advantage of this formulation is that it can be easily generalised to an arbitrary property, providing that a numerical sequence $\boldsymbol{\xi} \to \boldsymbol{s}$ can be established and sufficient numerical precision retained.

Detailed examples of the application of the $\boldsymbol{t}$-$\boldsymbol{s}$ formalism to the construction of a KEO for an XY_2 system are presented in Section 4.1.

3.6 JACOBI COORDINATES KEOs

In this section we illustrate the advantages of Jacobi coordinates (see Section 2.4.3) for the KEO of polyatomic molecules. Let us consider an example of the Jacobi frame of an A_2B_2-type molecule (e.g. acetylene C_2H_2) shown in Fig. 3.1. Here the z-axis is chosen along the A–A bond vector (a valence-type Jacobi vector) of the length R, while for the two B-atoms the Jacobi vectors originate from the centre-of-mass of the entire molecule. The advantage of this simplistic Jacobi frame is to describe the relative positions of atoms B_1 and B_2 using spherical polar coordinates r_i, θ_i and ϕ_i ($i = 1, 2$). Indeed the Cartesian coordinates of B_1 and B_2 are given simply by

$$\begin{aligned} x_1 &= r_1 \sin\theta_1 \cos\phi_1, & x_2 &= r_2 \sin\theta_2 \cos\phi_2, \\ y_1 &= r_1 \sin\theta_1 \sin\phi_1, & y_2 &= r_2 \sin\theta_2 \sin\phi_2, \\ z_1 &= r_1 \cos\theta_1, & z_2 &= r_2 \cos\theta_2, \end{aligned}$$

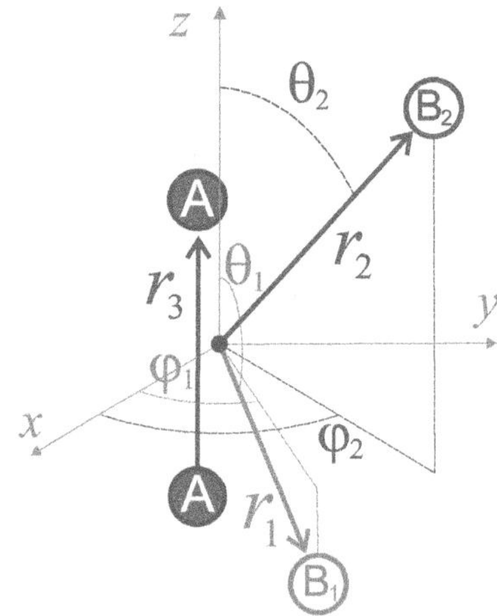

Figure 3.1 Jacobi coordinates for an A_2B_2-type molecule with two polyspherical coordinates r_i, θ_i and ϕ_i, $i = 1, 2$.

where we use the notation x_i, y_i, z_i in place of the Cartesian components $r_{i,x}, r_{i,y}, r_{i,z}$. The positions of the two A-atoms are selected to satisfy the centre-of-mass condition

$$m_{\mathrm{B}}(r_{1,\alpha} + r_{2,\alpha}) + m_{\mathrm{A}}(r_{3,\alpha} + r_{4,\alpha}) = 0, \quad \alpha = x, y, z$$

with the constraint for the A–A bond length:

$$\sum_{\alpha=x,y,z} (r_{3,\alpha} - r_{4,\alpha})^2 = r_3^2.$$

At this point, one can decide to work either in the $3N-6$ or $3N-5$ frame. For the latter, all seven coordinates $\boldsymbol{\xi} = \{R, r_1, \theta_1, \phi_1, r_2, \theta_2, \phi_2\}$ are used as vibrational coordinates, complemented by two Euler angles α and β to control the orientation of the z axis. For the former, the $3N-6=6$ coordinates are $\boldsymbol{\xi} = \{R, r_1, \theta_1, r_2, \theta_2, \Delta\phi\}$ with the azimuthal coordinate $\Delta\phi = \phi_2 - \phi_1$.

The polyspherical coordinates in either of these two choices have well-established properties. For example, their derivatives with respect to $r_{n,\alpha}$ which define the $\boldsymbol{s}$-vectors are given by the standard expression for the polar spherical coordinates as follows:

$$\begin{aligned}
\frac{\partial r_i}{\partial x_i} &= \sin\theta_i \cos\phi_i, & \frac{\partial \theta_i}{\partial x_i} &= \frac{\cos\theta_i \cos\phi_i}{r_i}, & \frac{\partial \phi_i}{\partial x_i} &= -\frac{\sin\phi_i}{r_i \sin\theta_i} \\
\frac{\partial r_i}{\partial y_i} &= \sin\theta_i \sin\phi_i, & \frac{\partial \theta_i}{\partial y_i} &= \frac{\cos\theta_i \sin\phi_i}{r_i}, & \frac{\partial \phi_i}{\partial y_i} &= \frac{\cos\phi_i}{r_i \sin\theta_i} \\
\frac{\partial r_i}{\partial z_i} &= \cos\theta_i, & \frac{\partial \theta_i}{\partial z_i} &= -\frac{\sin\theta_i}{r_i}, & \frac{\partial \phi_i}{\partial z_i} &= 0.
\end{aligned}$$

and

$$\frac{\partial \xi_\lambda}{\partial r_{i\alpha}} = s_{\lambda,i\alpha} \tag{3.56}$$

for ξ_λ as one of $\{r_i, \theta_i, \phi_i\}$.

One can also introduce angular momentum operators $\hat{L}_i = \boldsymbol{r}_i \times \frac{\partial}{\partial \boldsymbol{r}_i}$ and $\hat{L}_i^2$ associated with the Jacobi vectors $\boldsymbol{r}_1$, $\boldsymbol{r}_2$ and $\boldsymbol{r}_3$:

$$\begin{aligned}
\hat{L}_i &= i\hbar \begin{pmatrix} \sin\phi_i \frac{\partial}{\partial\theta_i} + \cot\theta_i \cos\phi_i \frac{\partial}{\partial\phi_i} \\ -\cos\phi_i \frac{\partial}{\partial\theta_i} + \cot\theta_i \sin\phi_i \frac{\partial}{\partial\phi_i} \\ -\frac{\partial}{\partial\phi_i} \end{pmatrix}, \\
\hat{L}_i^2 &= -\hbar^2 \left[\frac{1}{\sin\theta_i} \frac{\partial}{\partial\theta_i} \left(\sin\theta_i \frac{\partial}{\partial\theta_i} \right) + \frac{1}{\sin\theta_i^2} \frac{\partial^2}{\partial\phi_i^2} \right].
\end{aligned}$$

The KEO is then obtained via the standard Laplacian transformation:

$$\hat{T} = \sum_{i=1}^{3} \left[-\frac{\hbar^2}{2\mu_i r_i^2} \frac{\partial}{\partial r_i} r_i^2 \frac{\partial}{\partial r_i} + \frac{1}{2\mu_i r_i^2} \hat{L}_i^2 \right]$$

with the Euclidean integration volume element given by

$$d\tau = \prod_i^3 r_i^2 \sin\theta_i dr_i d\theta_i d\phi_i.$$

Here, the reduced masses μ are $\mu_1 = \mu_2 = m_\mathrm{A}$ and $\mu_3 = m_\mathrm{B}/2$.

It should be noted that $\hat{T}$ becomes singular for $\theta_i = 0°$ and $\theta_i = 180°$. However, with the right basis set (e.g. spherical harmonics), the matrix elements are all finite.

3.7 KEO AS A TAYLOR-TYPE EXPANSION

The on-the-fly numerical formulations of the $\boldsymbol{t}$–$\boldsymbol{s}$ and Sørensen formalisms at any instantaneous geometry are well suited to variational black-box algorithms, such as those based on the DVR or multi-dimensional Gaussian-quadrature integration methods, which require knowledge of the KEO for any instantaneous configurations of the nuclei for building the Hamiltonian matrix. In this case, the KEO obtained is effectively exact, while integration is subject to quadrature approximations.

In the following we present an alternative approach to the construction of the KEO, as a Taylor-type expansion, which belongs to another powerful and popular variational black-box methodology based on the finite basis-set representation (FBR) approach and is also well suited for purely numerical implementations. A Taylor form of the KEO has a sum-of-products form and therefore works especially well in conjunction with product-form basis sets. In this case, the KEO is approximate, while the evaluation of the matrix elements is nominally exact.

In presenting a Taylor-expanded KEO, we follow the same $\boldsymbol{t}$–$\boldsymbol{s}$ formalism (or its Sørensen derivative) with special emphasis on the compatibility with a numerically motivated, black-box implementation.

3.7.1 Recursive expansion scheme

In this section we build the KEO as a Taylor type expansion in terms of some arbitrary internal coordinates $\boldsymbol{\xi} = \{\xi_1, \xi_2, \dots, \xi_M\}$ ($M = 3N - 6$). Let us assume the elements of the $\boldsymbol{s}$-transformation matrix to have a Taylor form:

$$s_{n,i\alpha} = \sum_{i,j,k,l,\dots} s^{(n,i\alpha)}_{i,j,k,l,\dots} \xi_1^i \xi_2^j \xi_3^k \xi_4^l \cdots$$

expanded up to some $N_\mathrm{kin}^\mathrm{th}$ order in terms of the $\boldsymbol{\xi}$ internal (vibrational) coordinates. The KEO factors $\boldsymbol{G}$ are obtained as products of two $\boldsymbol{s}$-matrices (see Eq. (3.14)) and therefore will have a similar expansion form

$$G_{n,n'} = \sum_{i,j,k,l,\dots} G^{(n,n')}_{i,j,k,l,\dots} \xi_1^i \xi_2^j \xi_3^k \xi_4^l \cdots$$

In order to simplify manipulations of multi-dimensional the Taylor expansions, let us introduce the following compact multi-index representation for a Taylor-type expansion of $f(\boldsymbol{\xi})$ of order N:

$$f(\boldsymbol{\xi}) \approx \sum_{L=0}^{N} \sum_{L[\boldsymbol{l}]} f_{L[\boldsymbol{l}]} (\boldsymbol{\xi})^{L[\boldsymbol{l}]}$$

$$\equiv \sum_{L=0}^{N} \sum_{l_1=0}^{L} \sum_{l_2=0}^{(L-l_1)} \sum_{l_3=0}^{(L-l_1-l_2)} \sum_{l_{M-1}=0}^{(L-l_1-l_2-\ldots l_{M-2})} f^L_{l_1 l_2 l_3 \ldots l_M} \prod_i \xi_i^{l_i}, \tag{3.57}$$

where $L[\boldsymbol{l}]$ is a set of $\{L, l_1, l_2, l_3, \ldots, l_M\}$ constrained with $l_1 + l_2 + l_3 + \cdots l_M = L$ and $M = 3N - 6$. For each set of $\{L, l_1, l_2, \ldots l_M\}$, the index l_M given in Eq. (3.57) is redundant and set to

$$l_M = L - l_1 - l_2 - \ldots - l_{M-1}.$$

The expansion terms in Eq. (3.57) are arranged according to their total power L, which defines the perturbation orders $O(\epsilon^L)$ in the expansion of the KEO assuming that each ξ_λ is a small displacement from equilibrium of some order of magnitude $\sim O(\epsilon)$.

We will also employ the Leibnitz product rule when forming a product of two series. Namely, a product of two series $\sum_{L[\boldsymbol{l}]} f_{L[\boldsymbol{l}]}(\boldsymbol{\xi})^{L[\boldsymbol{l}]}$ and $\sum_{L[\boldsymbol{l}]} g_{L[\boldsymbol{l}]}(\boldsymbol{\xi})^{L[\boldsymbol{l}]}$ is a series of the type $\sum_{L[\boldsymbol{l}]} h_{L[\boldsymbol{l}]}(\boldsymbol{\xi})^{L[\boldsymbol{l}]}$ with the expansion coefficients defined as

$$h_{L[\boldsymbol{l}]} = \sum_{K=0}^{L} \sum_{K[\boldsymbol{k}]} f_{(L-K)[\boldsymbol{l}-\boldsymbol{k}]}\, g_{K[\boldsymbol{k}]}. \tag{3.58}$$

In order to simplify the manipulation of the products of the Taylor series, we introduce the following truncation rule: the order of the expansion h is taken as $\mathbf{min}\{\text{order}(f), \text{order}(g)\}$.

The derivative $\partial/\partial\xi_n$ of the polynomial f of order $L+1$ is a polynomial h of order L with the coefficients

$$h^L_{l_1, l_2, \ldots, l_n, \ldots, l_{3N-6}} = (l_n + 1) f^{L+1}_{l_1, l_2, \ldots, l_n+1, \ldots, l_M}, \tag{3.59}$$

which in the compact form of Eq. (3.57) is given by

$$h_{L[\boldsymbol{l}]} = (l_n + 1) f_{(L+1)[\boldsymbol{l_n}+\mathbf{1}]}. \tag{3.60}$$

Here $(L+1)[\boldsymbol{l_n} + \mathbf{1}]$ is a set of the total power $L+1$ and the nth index increased by 1:

$$(L+1)[\boldsymbol{l_n} + \mathbf{1}] \equiv l_1, l_2, \ldots, l_n + 1, \ldots, l_M, \quad l_1 + l_2 + l_3 + \ldots l_M = L.$$

We start the construction of the KEO by expanding both $\boldsymbol{s}_{\lambda,i}$ and $\boldsymbol{t}_{i,\mu}$ as power series in ξ_n

$$s_{\lambda,i\alpha}(\boldsymbol{\xi}) = \sum_{L=0}^{N}\sum_{L[\boldsymbol{l}]} s_{L[\boldsymbol{l}]}^{\lambda,i\alpha}\,(\boldsymbol{\xi})^{L[\boldsymbol{l}]}, \tag{3.61}$$

$$t_{i\alpha,\mu}(\boldsymbol{\xi}) = \sum_{L=0}^{N}\sum_{L[\boldsymbol{l}]} t_{L[\boldsymbol{l}]}^{i\alpha,\mu}\,(\boldsymbol{\xi})^{L[\boldsymbol{l}]}. \tag{3.62}$$

Substituting Eqs. (3.61) and (3.62) into the orthogonality condition of Eq. (3.11),

$$\sum_{n\alpha} s_{i,n\alpha} t_{n\alpha,i'} = \delta_{i,i'},$$

and collecting coefficients of $\xi_1^{l_1}\xi_2^{l_2}\xi_3^{l_3}\ldots$ of order $L = l_1 + l_2 + \ldots l_M$, we obtain systems of linear equations for $s_{L[\boldsymbol{l}]}^{\lambda,i\alpha}$, which are defined recursively as

$$\sum_{i\alpha} t_{0[0]}^{i\alpha,\mu}\, s_{0[0]}^{\lambda,i\alpha} = \delta_{\lambda,\mu}, \tag{3.63}$$

$$\sum_{i\alpha} t_{0[0]}^{i\alpha,\mu}\, s_{L[\boldsymbol{l}]}^{\lambda,i\alpha} = b_{L[\boldsymbol{l}]}^{\mu,\lambda}, \quad L = 1\ldots N, \tag{3.64}$$

where

$$b_{L[\boldsymbol{l}]}^{\mu,\lambda} = -\sum_{K=0}^{L-1}\sum_{K[\boldsymbol{k}]}\sum_{i\beta} t_{(L-K)[\boldsymbol{l}-\boldsymbol{k}]}^{i\beta,\mu}\, s_{K[\boldsymbol{k}]}^{\lambda,i\beta}. \tag{3.65}$$

For each combination of indices $(L[\boldsymbol{l}], \lambda, \mu)$, Eqs. (3.63) and (3.64) represent a set of linear equations of type $\boldsymbol{T}\boldsymbol{x} = \boldsymbol{b}$ for the vector $\boldsymbol{x}$ with elements $x_{i\alpha} = s_{L[\boldsymbol{l}]}^{\lambda,i\alpha}$ $(i = 1\ldots N; \alpha = x, y, z; \lambda = 1\ldots 3N)$. The $3N$-dimensional 'right-hand-side' vector $\boldsymbol{b}$ with the elements $b_\mu = b_{L[\boldsymbol{l}]}^{\mu,\lambda}$ is recursively determined at each perturbation order L by means of Eq. (3.65). The corresponding $3N \times 3N$ $\boldsymbol{T}$-matrix is given by the elements $T_{\mu,i\alpha} = t_{0[0]}^{i\alpha,\mu}$ $(\mu = 1\ldots 3N)$, which are independent of λ. The linear equation systems in Eqs. (3.63, 3.64) can be prepared and solved numerically. The order of the system is small $(3N)$, i.e. any appropriate linear algebra routine from a standard computer library should be capable of obtaining $s_{L[\boldsymbol{l}]}^{\lambda,i\alpha}$.

The $\boldsymbol{t}_{i,\mu}$ coefficients required for building the linear system in Eqs. (3.63–3.65) are constructed by expanding the definition of the $\boldsymbol{t}$-matrix given in Eq. (3.26). The $\boldsymbol{t}_{i,\mu}$ vectors are linear combinations of the Cartesian body-frame position vectors $\boldsymbol{r}_i$ or their partial derivatives $\partial r_{i\alpha}/\partial\xi_n$, which we present as power series in the vibrational variables ξ_n using the notation of Eq. (3.57)

$$r_{i\alpha}(\boldsymbol{\xi}) = \sum_{L\geq 0}\sum_{L[\boldsymbol{l}]} R_{L[\boldsymbol{l}]}^{i\alpha}\,(\boldsymbol{\xi})^{L[\boldsymbol{l}]}, \tag{3.66}$$

$$\frac{\partial r_{i\alpha}}{\partial\xi_n} = \sum_{L\geq 0}\sum_{L[\boldsymbol{l}]} R_{L[\boldsymbol{l}]}^{i\alpha,n}\,(\boldsymbol{\xi})^{L[\boldsymbol{l}]}, \tag{3.67}$$

where we assume that all necessary derivatives of $r_{i\alpha}$ with respect to ξ_n are defined. Insertion into Eqs. (3.26) and (3.62) yields

$$t^{i\alpha,\alpha'}_{L[\boldsymbol{l}]} = \delta_{\alpha,\alpha'}\delta_{L,0}, \tag{3.68}$$

$$t^{i\alpha,\beta}_{L[\boldsymbol{l}]} = \sum_{\gamma} \epsilon_{\alpha\beta\gamma} R^{i\gamma}_{L[\boldsymbol{l}]}, \tag{3.69}$$

$$t^{i\alpha,n}_{L[\boldsymbol{l}]} = R^{i\alpha,n}_{L[\boldsymbol{l}]}. \tag{3.70}$$

In our scheme, all quantities are given as polynomials in the ξ_n variables. This includes the kinetic energy terms $G_{\lambda,\lambda'}$ and the pseudo-potential function U in Eqs. (3.14) and (3.19), respectively: $G_{\lambda,\lambda'}$ is a product of two polynomials $s_{\lambda,i\beta}$ and $s_{\lambda',i\beta}$ expressed as power series in the ξ_n variables:

$$G_{\lambda,\lambda'}(\boldsymbol{\xi}) = \sum_{L=0}^{N_{\text{kin}}} \sum_{L[\boldsymbol{l}]} G^{\lambda,\lambda'}_{L[\boldsymbol{l}]} (\boldsymbol{\xi})^{L[\boldsymbol{l}]} \tag{3.71}$$

with expansion coefficients defined by

$$G^{\lambda,\lambda'}_{L[\boldsymbol{l}]} = \sum_{i=1}^{N} \sum_{\beta=x,y,z} \frac{1}{m_i} \sum_{K=0}^{L} \sum_{K[\boldsymbol{k}]} s^{\lambda,i\beta}_{(L-K)[\boldsymbol{l}-\boldsymbol{k}]} s^{\lambda',i\beta}_{K[\boldsymbol{k}]}. \tag{3.72}$$

The expansion of the pseudo-potential function U

$$U(\xi) = \sum_{L=0}^{N_{\text{kin}}} \sum_{L[\boldsymbol{l}]} U_{L[\boldsymbol{l}]} (\boldsymbol{\xi})^{L[\boldsymbol{l}]} \tag{3.73}$$

is obtained using Eq. (3.19) in the same manner by applying operations of Eqs. (3.58) and (3.60) to the $\boldsymbol{s}$ expansion in Eq. (3.61).

The construction of the kinetic energy coefficients $G_{\lambda,\lambda'}$ and the pseudo-potential U by means of the recurrence relations in Eqs. (3.63) and (3.64) is designed for purely numerical implementations. The kinetic energy coefficients $G_{\lambda,\lambda'}$ in Eq. (3.71) and the pseudo-potential function U in Eq. (3.73) are expected to be truncated after the same perturbation order $L = N_{\text{kin}}$. Since the pseudo-potential term series is constructed from first and second derivatives of the series of the $\boldsymbol{s}_{\lambda,n}$ vectors, the latter have to be originally expanded at least up to order $N_{\text{kin}} + 2$. A KEO expansion in a typical polyatomic molecule application is expected to have N_{kin} of 4 to 8.

The algebra of operations with series introduced in this section is completely generic, i.e. applicable for any size of molecule (at least in principle), and therefore well suited for black-box numerical algorithms. For example, a mapping of the expansion coefficients $f^{L}_{l_1,l_2,\ldots,l_M}$ for a polynomial of dimension M and expansion order $L_{\max}$ to a vector $f[i]$ $(i = 1, \ldots i_{\max})$

$$f_{l_1,l_2,\ldots,l_M} \to f[i]$$

can be controlled by the 2D integer array $N[i,m] = l_m$, where m is the mode count ($m = 1 \dots M$) and i is the index defining the 1D parameter array $f[i]$ assuming the following form:

$$F(\boldsymbol{\xi}) = \sum_i f[i]\, \xi_1^{N[i,1]} \xi_1^{N[i,2]} \dots \xi_M^{N[i,M]}. \tag{3.74}$$

The total number of polynomial coefficients given in Eq. (3.57), i.e. the size of the 1D array $F[j]$, is given by

$$i_{\max} = \frac{(N+1)(N+2)\dots(N+M)}{M!}.$$

The mapping $i \to \{l_1, l_2, \dots, l_M\}$ can be implemented as a compact recursive procedure, such as in the following pseudo-code example INDEX returning the 2D mapping array $\boldsymbol{N} = \{N[i,m]\}$ for the dimension M and expansion order $L_{\max}$: this protocol is summarised in the following pseudo-code:

```
procedure INDEX(N,m,M,N_max,i)
    N[1 : i_max, 1 : M] ← 0
    i ← 1
    for L = 0, N_max do
        while m < M − 1 do
            i' ← 0
            while i' ≤ i_max do
                if m = M − 1 then
                    if Σ_i' N[i', m] ≤ L then
                        i ← i + 1
                        N[i, M] = L − Σ_{m'=1}^{M−1} N[i, m']
                    end if
                else
                    CALL INDEX(N,m + 1,M,N_max-1,i)
                end if
            end while
            N[i, m] ← N[i, m] + 1
        end while
    end forreturn
end procedureN
```

Then the full mapping of $\boldsymbol{N} \to \{l_1, l_2, \dots, l_M\}$ is generated by calling INDEX($\boldsymbol{N}, m = 1, M, i = 0$). Recursive structures have the advantage of defining dynamically included loops for arbitrary dimensions, which is important for black-box algorithms. The inverse mapping $F_{l_1,l_2,l_3 \dots l_M} \to f[j]$ is less efficient to maintain as a generalised object of arbitrary dimension. Although a on-the-fly reconstruction of the index i for a given set $\{l_1, l_2, l_3 \dots l_M\}$ is possible with the same procedure INDEX, it is inefficient. The good news, however, is that only the direct mapping

$N[i,m] \to l_m$ is actually needed for the representation of polynomial in the form given in Eq. (3.74).

3.7.2 Taylor-expanded s-vectors in the linearised coordinates using the t–s algorithm and its Sørensen's derivative

Let us apply the Taylor expansion approach to the linearised coordinates representation. As introduced earlier in Section 2.5, the linearised (rectilinear) coordinates $\xi_\lambda^{\rm lin}$ are linked to the Cartesian $\boldsymbol{r}_n$ coordinates via the linear transformations

$$\begin{aligned} (r_{i\alpha} - r_{i\alpha}^{\rm e}) &= \sum_{\lambda=1}^{M} A_{\lambda,i\alpha}\xi_\lambda^{\rm lin}, \\ \xi_\lambda^{\rm lin} &= \sum_{i=1}^{N} \sum_{\alpha=x,y,z} B_{\lambda,i\alpha}(r_{i\alpha} - r_{i\alpha}^{\rm e}). \end{aligned} \tag{3.75}$$

The elements of the linearised $\boldsymbol{B}$-matrix here are defined as the first derivatives of the valence coordinates with respect to the Cartesian positions $r_{i\alpha}$ taken at equilibrium, which is the same as the equilibrium vibrational elements of the $\boldsymbol{s}$-matrix given in Eq. (3.21):

$$B_{\lambda,i\alpha} = s_{\lambda,i\alpha}^{\rm (vib)e}. \tag{3.76}$$

In turn, the elements $A_{n,i\alpha}$ are obtained by inverting the matrix $\boldsymbol{b}$ in Eqs. (2.64–2.66), which can be formulated and solved numerically.

Comparing Eq. (3.75) and its first derivative

$$\frac{\partial r_{i\alpha}(\boldsymbol{\xi})}{\partial \xi_n} = A_{n,i\alpha}$$

with the corresponding Taylor expansions in Eqs. (3.66) and (3.67), for the corresponding expansion coefficients $R_{L[\boldsymbol{l}]}^{i\alpha}$ we obtain

$$\begin{aligned} R_{0[0]}^{i\alpha} &= r_{i\alpha}^{\rm e}, \\ R_{1[l_n=1]}^{i\alpha} &= A_{n,i\alpha}, \quad 1[l_n=1] \equiv \{0,0,\ldots,1,\ldots,0,0\}, \\ R_{0[0]}^{i\alpha,n} &= A_{n,i\alpha}. \end{aligned}$$

The elements of the $\boldsymbol{t}$-matrix given in Eq. (3.26) in this case are trivial linear functions of $\xi_\lambda^{\rm lin}$:

$$\begin{aligned} t_{i\alpha,\alpha'}^{\rm tr} &= \delta_{\alpha,\alpha'}, \\ t_{i\alpha,\beta}^{\rm rot} &= \textstyle\sum_\gamma \epsilon_{\alpha\beta\gamma}(r_{i\gamma}^{\rm e} + \sum_\lambda A_{\lambda,i\gamma}\xi_\lambda^{\rm lin}), \\ t_{i\alpha,\lambda}^{\rm vib} &= A_{\lambda,i\alpha}, \end{aligned} \tag{3.77}$$

with the non-zero expansion coefficients $t^{i\alpha,\mu}_{L[\boldsymbol{l}]}$ in Eqs. (3.68–3.70) given by

$$t^{i\alpha,\alpha'}_{L[\boldsymbol{l}]} = \delta_{\alpha,\alpha'}\delta_{L,0}, \tag{3.78}$$

$$t^{i\alpha,g}_{0[0]} = \sum_{\gamma} \epsilon_{\alpha g\gamma} r^{\mathrm{e}}_{i\gamma}, \tag{3.79}$$

$$t^{i\alpha,g}_{1[l_\lambda=1]} = \sum_{\gamma} \epsilon_{\alpha g\gamma} A_{\lambda,i\gamma}, \tag{3.80}$$

$$t^{i\alpha,n}_{0[0]} = A_{n,i\alpha}. \tag{3.81}$$

This fully defines the recursive linear equations in Eqs. (3.63–3.65) for the expansion coefficients $s^{\lambda,i\alpha}_{L[\boldsymbol{l}]}$ as a generalised algorithm suitable for pure numerical applications.

Let us now consider Sørensen's formulation of the $\boldsymbol{t}$–$\boldsymbol{s}$ algorithm for the case of the Taylor-expanded linearised representation using the $\boldsymbol{C}$-vector formalism (Section 3.4). This formalism helps to reduce the size of the inversion problem to the rotational part of the $\boldsymbol{s}$-vectors by decoupling it from the vibrational part.

A Sørensen $\boldsymbol{C}$-vector for the three Eckart conditions is defined in Eq. (3.44). Using it together with Eq. (3.26) for the $\boldsymbol{J}$-matrix given in Eq. (3.42), we obtain

$$J_{gg'} = I^{\mathrm{e}}_g \delta_{g,g'} + \sum_{n=1}^{N} m_n (\boldsymbol{e}_g \times \boldsymbol{r}^{\mathrm{e}}_n) \cdot (\boldsymbol{e}_{g'} \times \Delta\boldsymbol{r}_n), \tag{3.82}$$

where I^{e}_g are the equilibrium moments of inertia and $\boldsymbol{e}_\alpha$ are three xyz unit vectors $\boldsymbol{e}_x$, $\boldsymbol{e}_y$ and $\boldsymbol{e}_z$. One can see that $J_{gg'}$ in Eq. (3.82) has the same form as the normal-mode version of $\boldsymbol{J}$ in Eq. (3.33).

Eq. (3.82) can be simplified using the scalar quadruple product to obtain

$$J_{gg'} = I^{\mathrm{e}}_g \delta_{g,g'} + \sum_{i} m_i (\delta_{g,g'} \boldsymbol{r}^{\mathrm{e}}_i \cdot \Delta\boldsymbol{r}_i - r^{\mathrm{e}}_{ig} \Delta r_{ig'}),$$

where $\Delta r_{i\alpha}$ is given by the linear expansion in Eq. (3.75). Thus $J_{gg'}$ is linear in $\boldsymbol{\xi}^{\mathrm{lin}}$:

$$J_{gg'} = J^{(0)}_{gg'} + \sum_{\lambda} J^{gg'}_{\lambda} \xi^{\ell}_{\lambda}, \tag{3.83}$$

where

$$J^{(0)}_{gg'} = I^{\mathrm{e}}_g \delta_{g,g'}.$$

The matrix $\boldsymbol{\eta}$ as $\boldsymbol{J}^{-1}$ (see Eq. (3.43)) can also be found as a Taylor expansion

$$\eta_{gg'} = \sum_{i,j,k,\ldots} \eta^{gg'}_{i,j,k,\ldots} \xi^i_1 \xi^j_2 \xi^k_3 \ldots = \sum_{L=0}^{N} \sum_{L[\boldsymbol{l}]} \eta^{gg'}_{L[\boldsymbol{l}]} (\boldsymbol{\xi})^{L[\boldsymbol{l}]} \tag{3.84}$$

by solving Eq. (3.43) recursively for the coefficients $\eta^{gg'}_{i,j,k,\dots}$ as follows (here we omit the superscript 'lin' for simplicity and also use the compact summation form given in Eq. (3.57)). The zero-order coefficients $\boldsymbol{\eta}_{0[0]}$ are diagonal and are given by

$$\eta^{gg'}_{0[0]} \equiv \eta^{(0)}_{gg'} = \frac{1}{I^{\mathrm{e}}_{g}} \delta_{g,g'}.$$

All other coefficients $\boldsymbol{\eta}_{l_1,l_2,l_3,\dots}$ for $L = l_1 + l_2 + l_3 + \dots > 0$ are obtained by solving a recursive system of linear equations

$$\boldsymbol{\eta}^{(L+1)}_{l_1,l_2,\dots,l_\lambda+1,\dots} \boldsymbol{J}^{(0)} = -\boldsymbol{\eta}^{(L)}_{l_1,l_2,l_3\dots} \boldsymbol{J}_\lambda,$$

which in a compact form are also given by

$$\boldsymbol{\eta}_{(L+1)[l_\lambda+1]} \boldsymbol{J}^{(0)} = -\boldsymbol{\eta}_{(L)[\boldsymbol{l}]} \boldsymbol{J}_\lambda$$

with right hand sides determined at the $(L-1)$th iteration.

The rotational matrix $\boldsymbol{s}^{(\mathrm{rot})}$ is then obtained via Eq. (3.41) as a Taylor expansion:

$$\boldsymbol{s}^{(\mathrm{rot})}_{g,i} = \sum_{L=0}^{N_{\mathrm{kin}}} \sum_{L[\boldsymbol{l}]} \boldsymbol{s}^{g,i}_{L[\boldsymbol{l}]} (\boldsymbol{\xi})^{L[\boldsymbol{l}]}.$$

The vibrational part $\boldsymbol{s}^{(\mathrm{vib})}$, obtained as a solution of the vibrational part of Eq. (2.61), is then given by

$$s^{\mathrm{vib}}_{\lambda,i\alpha} = s^{\mathrm{e}}_{\lambda,i\alpha} - \sum_{g,\lambda'} \xi^{\mathrm{lin}}_{\lambda'} \zeta^{g}_{\lambda',\lambda} s^{\mathrm{rot}}_{g,i\alpha}, \tag{3.85}$$

which makes use of the rotational vector $\boldsymbol{s}^{\mathrm{rot}}_{g,i}$. The $\zeta^{g}_{\lambda',\lambda}$ are the Coriolis coefficients

$$\zeta^{g}_{\lambda',\lambda} = \sum_{i\alpha,\beta} \epsilon_{g\alpha\beta} A_{i\alpha,\lambda'} B_{\lambda,i\beta} \tag{3.86}$$

or in the vector notation

$$\zeta^{g}_{\lambda',\lambda} = \sum_{i} \boldsymbol{e}_g \cdot (\boldsymbol{A}_{i,\lambda'} \times \boldsymbol{B}_{\lambda,i}).$$

The latter is consistent with the normal-mode coefficients in Eq. (3.31). The expansion coefficients of the kinetic energy factors $\boldsymbol{G}$ are then calculated using Eq. (3.72).

An interesting observation taken from Eq. (3.85) is that the linearised coordinates vector $\boldsymbol{s}^{\mathrm{vib}}_{\lambda,i}$ are not simple derivatives of the internal coordinates $\boldsymbol{\xi}^{\mathrm{lin}}$ with respect to the Cartesian displacements $\Delta\boldsymbol{R}_n$, which differs from the definition of the valence coordinates $\boldsymbol{s}^{\mathrm{vib}}$ matrix. This is a consequence of the dependence of the linearised coordinates on the orientation of the body-fixed frame via the corresponding rotational constraints.

Formally, the vibrational angular momenta $\hat{\pi}_\alpha$ defined in Eq. (3.29) are not part of the KEO construction, but they can be useful in different applications. For example, the square of the total angular momentum $\hat{L}^2 = \{\hat{\pi}_x^2, \hat{\pi}_y^2, \hat{\pi}_z^2\}$ can be used to build vibration basis set functions as eigenfunctions of the vibrational angular momentum quantum number l as follows:

$$\hat{L}^2|v, l\rangle = \hbar^2 l^2 |v, l\rangle,$$

where v is some generic vibrational quantum number and we use $\hat{L}$ instead of $\hat{\pi}$ for consistency. For example, wavefunctions of a 2D isotropic harmonic oscillator are commonly classified in accordance with l; see Section 5.3.2. For these purposes, we can represent $\hat{L}^2$ as a quadratic form given by

$$\hat{L}^2 = \hbar^2 \sum_{\mu,\mu'} \frac{\partial}{\partial \xi_\mu} L_{\mu,\mu'}(\boldsymbol{\xi}) \frac{\partial}{\partial \xi'_\mu}, \tag{3.87}$$

where the factors $L_{\mu,\mu'}(\boldsymbol{\xi})$ are quadratic functions of the rectilinear coordinates $\boldsymbol{\xi}$:

$$L_{\mu,\mu'}(\boldsymbol{\xi}) = \sum_g \sum_{\lambda,\lambda'} \zeta^g_{\lambda,\mu} \zeta^g_{\lambda',\mu'} \xi_\lambda \xi_{\lambda'}.$$

3.7.3 Non-rigid reference configuration and linearised coordinates

The representation of the KEO as a Taylor series, especially in terms of the rectilinear (linearised) coordinates, becomes inaccurate and even unusable, when a molecule exhibits non-rigid motion, including motion between multiple local minima with low-energy barriers between them. In order to remedy this flaw, at least for cases with a single flexible mode, a non-rigid version of the expansion-type KEO operators was proposed by Hougen, Bunker and Johns (1970) as the so-called non-rigid reference frame.

The method, which is commonly referred to as HBJ, consists of (1) expanding the KEO (and also the potential energy function) around a non-rigid, 1D reference configuration ('rigid bender') instead of a rigid equilibrium geometry and (2) treating the corresponding non-rigid degree of freedom, say ρ, as part of the rotational modes.

For example, in the case of a triatomic molecule in the valence coordinate representation, one can treat the bending mode as a non-rigid degree of freedom. For reasons that will become clear later, in this case, the non-rigid coordinate is the bending mode which is described by the angle $\rho = 180° - \alpha$ (see Fig. 3.2). We then seek the KEO factors $\boldsymbol{G}$ and U in the following sum-of-products form:

$$G_{ij} = \sum_{k_1,k_2} g^{(ij)}_{k_1,k_2}(\rho) \xi_1^{k_1} \xi_2^{k_2},$$

$$U = \sum_{k_1,k_2} u_{k_1,k_2}(\rho) \xi_1^{k1} \xi_2^{k2}, \tag{3.88}$$

where the two stretching modes $\xi_1 = \Delta r_1$ and $\xi_2 = \Delta r_2$ represent the vibrational degrees of freedom. According to HBJ, the non-rigid degree of freedom is combined with the three rotational modes (Euler angles) into a 4D rotation-torsion set. In practical numerical applications, the 1D functions $g^{(ij)}_{k_1,k_2}(\rho)$ and $u(\rho)$ are defined on a grid ρ_k. In the following we will show how the $\boldsymbol{t}$–$\boldsymbol{s}$-formalism can be modified for the non-rigid case.

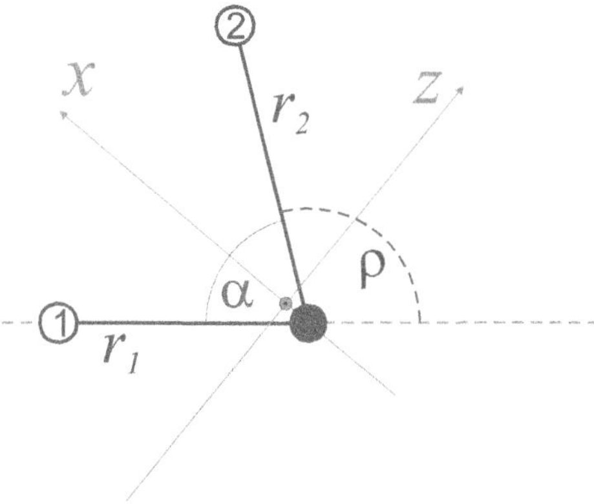

Figure 3.2 A non-rigid configuration of XY_2 described by the bending coordinate $\rho = \pi - \alpha$.

We start by defining the non-rigid reference configuration as the ρ-dependent body-fixed Cartesian coordinates $a_{n\alpha}(\rho) = r^{\rm ref}_{n\alpha}(\rho)$ ($n = 1, \ldots, N$; $\alpha = x, y, z$). The instantaneous reference geometry plays the role of an 'equilibrium' configuration for the given value of ρ and is constructed to satisfy (i) the centre-of-mass condition and (ii) the PAS condition (our method of choice here).

Alternatively, the PAS condition can be replaced by the following differential condition designed to reduce the KEO coupling between the vibration and torsion of the molecule (Hougen et al., 1970):

$$\sum_{n=1}^{N} m_n \boldsymbol{r}_n^{\rm NR} \times \frac{\partial \boldsymbol{r}_n^{\rm NR}}{\partial \rho} = 0, \tag{3.89}$$

which requires that the angular momentum of the reference configuration vanishes in the molecule-fixed axis system. For a triatomic molecule, this differential equation assumes an analytic solution; otherwise it can be solved numerically. In order to satisfy this criterium, the non-rigid degree of freedom (ρ in this case) comes out as a solution of Eq. (3.89) as part of the definition of the non-rigid reference configuration. Another alternative to PAS as the reference system is when the molecular non-rigid frame can be chosen to follow the minimum energy path obtained by optimising the molecular geometry along the non-rigid degree of freedom as a so-called reaction path. This choice helps to simplify the representation of the potential energy function in contrast to the optimisation of the KEO given in Eq. (3.89).

In order to construct the coordinate transformation to the BF frame, the $\boldsymbol{t}$–$\boldsymbol{s}$ formalism is modified by moving the non-rigid coordinate ρ into a separate category and treating it more like a rotational degree of freedom. The $\boldsymbol{t}$-matrix in Eq. (3.26) for the non-rigid (NR) reference form is then given by

$$t_{n\alpha,\alpha'} = \delta_{\alpha,\alpha'}, \qquad \text{(translation)} \tag{3.90}$$

$$t_{n\alpha,\beta} = \sum_{\gamma} \epsilon_{\alpha\beta\gamma} r_{n\gamma}^{\mathrm{NR}}, \qquad \text{(rotation)} \tag{3.91}$$

$$t_{n\alpha,\rho} = \frac{\partial r_{n\alpha}^{\mathrm{NR}}}{\partial \rho}, \qquad \text{(non-rigid mode)} \tag{3.92}$$

$$t_{n\alpha,\lambda} = \frac{\partial r_{n\alpha}^{\mathrm{NR}}}{\partial \xi_{\lambda}}, \qquad \text{(vibration)} \tag{3.93}$$

where the vibrational part consists of $3N-7$ degrees of freedom. The pure rotational part is defined as usual via, e.g., the Eckart conditions. Similarly to Eckart's idea of decoupling the rotational and vibrational degrees of freedom, the formulation of the non-rigid term part is designed to minimise the coupling between the non-rigid mode ρ and 'rigid' vibrational degrees of freedom as given by

$$\sum_{m} m_n \frac{\partial \boldsymbol{r}_n^{\mathrm{NR}}}{\partial \rho} \cdot (\boldsymbol{r}_n - \boldsymbol{r}_n^{\mathrm{NR}}) = 0. \tag{3.94}$$

This is known as the Sayvetz condition (Sayvetz, 1939). Combining it with $\boldsymbol{t}_{n,\rho}$ in Eq. (3.92), we obtain the 4th constraint describing the non-rigid degree of freedom (in addition to the translational, rotational and vibrational constraints):

$$\sum_{m} m_n \boldsymbol{t}_{n,\rho}^{\mathrm{NR}} \cdot (\boldsymbol{r}_n - \boldsymbol{r}_n^{\mathrm{NR}}) = 0. \tag{3.95}$$

In the case of Sørensen's alternative for the $\boldsymbol{t}$–$\boldsymbol{s}$ formalism, the corresponding $\boldsymbol{C}$-vector is obtained using Eq. (3.95) (compare to Eckart's $\boldsymbol{C}$-vector in Eq. (3.44)):

$$\boldsymbol{c}_{\rho,n} = m_n \boldsymbol{t}_{n,\rho}^{\mathrm{NR}}.$$

Following the non-rigid alternative of the $\boldsymbol{C}$-vector Sørensen approach, we now form a 4×4 $\boldsymbol{J}$ 'rotational' matrix:

$$J_{g,g'} = \sum_{n} m_n \boldsymbol{t}_{n,g}^{\mathrm{NR}} \cdot \boldsymbol{t}_{n,g'}, \quad g = x, y, z, \rho,$$

which can be inverted at each value of ρ in Eq. (3.43) to get $\boldsymbol{\eta} = \boldsymbol{J}^{-1}$. The extended 4D rotational-torsional vectors $\boldsymbol{s}_{g,n}^{\mathrm{rot-tor}}$ $(g = x, y, z, \rho)$ are then given by (compare to the 3D rotational version in Eq. (3.41))

$$s_{g,n\alpha}^{\mathrm{rot-tor}}(\rho) = \sum_{g'} \eta_{gg'}(\rho)\, m_n\, t_{n\alpha,g'}^{\mathrm{NR}}(\rho), \quad \alpha = x, y, z. \tag{3.96}$$

The vibrational $\boldsymbol{s}^{(\mathrm{vib})}_{\lambda,n}$-vector ($\lambda = 1 \ldots 3N-7$) is defined as before using the $3N-7$ vibrational coordinates $\boldsymbol{\xi}$.

As an example, let us consider a linearised coordinates frame. Here we define the body-fixed Cartesian displacements as linear combinations of $3N-7$ linearised coordinates ξ^{ℓ}_{λ} around a non-rigid reference configuration $r^{\mathrm{NR}}_{n\alpha}(\rho)$ as given by

$$\begin{aligned} \boldsymbol{r}_n &= \boldsymbol{r}^{\mathrm{NR}}_n(\rho) + \sum_{\lambda} \boldsymbol{A}_{n,\lambda}(\rho)\xi^{\mathrm{lin}}_{\lambda}, \\ \xi^{\mathrm{lin}}_{\lambda} &= \sum_{n} \boldsymbol{B}_{\lambda,n}(\rho) \cdot (\boldsymbol{r}_n - \boldsymbol{r}^{\mathrm{NR}}_n(\rho)). \end{aligned}$$

The vectors $\boldsymbol{A}_{n,\lambda}(\rho)$ are obtained by satisfying the same conditions as before: (i) the centre-of-mass, (ii) PAS or Eq. (3.89) and (iii) orthogonality with $\boldsymbol{B}$,

$$\sum_{n} \boldsymbol{B}_{\lambda,n}\boldsymbol{A}_{n,\lambda'} = \delta_{\lambda\lambda'},$$

where the elements of the $\boldsymbol{B}$-matrix are first derivatives of $\boldsymbol{\xi}$ at each value of the non-rigid coordinate ρ:

$$B_{\lambda,n\alpha} = \left.\frac{\partial \xi_{\lambda}}{\partial r_{n\alpha}}\right|_{\boldsymbol{\xi}=\boldsymbol{\xi}^{\mathrm{NR}}}. \tag{3.97}$$

The vibrational $\boldsymbol{s}^{(\mathrm{vib})}_{\lambda,n}$ vector is defined analogously to Eq. (3.85):

$$s^{\mathrm{vib}}_{\lambda,n\alpha} = s^{\mathrm{NR}}_{\lambda,n\alpha}(\rho) - \sum_{g,\lambda'} \xi^{\mathrm{lin}}_{\lambda'}\zeta^{g}_{\lambda',\lambda}(\rho)s^{\mathrm{rot}}_{g,n\alpha}(\rho), \tag{3.98}$$

where $\zeta^{g}_{\lambda',\lambda}(\rho)$ are the ρ-dependent Coriolis coefficients

$$\zeta^{g}_{\lambda',\lambda}(\rho) = \sum_{n=1}^{N} \left(\frac{\partial \boldsymbol{A}_{n,\lambda'}(\rho)}{\partial \rho} \cdot \boldsymbol{B}_{\lambda,n}(\rho)\right) \tag{3.99}$$

and

$$\boldsymbol{B}_{\lambda,n}(\rho) = \boldsymbol{s}^{\mathrm{NR}}_{\lambda,n}(\rho). \tag{3.100}$$

In practical, numerical applications, a non-rigid reference KEO is generated on a grid of ρ_k points (e.g. equidistant or Gaussian-quadrature abscissas) basically following the same approach as for a rigid molecule with modifications for the torsional constraint. The $\boldsymbol{t}$-matrix is given by Eqs. (3.90–3.93), the $4D$ rot-torsional matrix $\boldsymbol{s}^{\mathrm{rot-tor}}(\rho_k)$ is obtained using Eq. (3.96) and the vectors $\boldsymbol{s}^{\mathrm{vib}}_{\lambda,n}(\rho_k)$ ($\lambda = 1..3N-7$) are given by Eq. (3.98). The kinetic energy factors $\boldsymbol{G}$ and U are contracted from the $\boldsymbol{s}$-matrix using the same equations (3.14) and (3.15). The latter requires derivatives of $\boldsymbol{s}(\rho_k)$ with respect to ρ, which can be evaluated using finite differences, cubic splines, etc.

It is important to note that throughout this chapter, the masses of electrons were completely ignored. This is an approximation which leads to a simpler form of the KEO. A proper treatment of the non-Born-Oppenheimer effects is beyond the subject of this book; however, it is worth mentioning that the contribution from electronic masses can be partly recovered by using atomic masses instead of nuclear ones in the expression of KEO.

MATERIALS USED

KEOs were derived in Wilson et al. (1955), Polo (1956), Watson (1968), Meyer and Günthard (1968), Watson (1970), Sørensen (1979), Jensen (1988), Carter and Handy (1986), Bramley et al. (1991), Chapuisat et al. (1991), Lukka (1995), Schwenke (1996), Bunker and Jensen (1998), Gatti et al. (1998), Leforestier et al. (2001), Gatti et al. (2001), Lauvergnat and Nauts (2002), Schwenke (2003), Watson (2004), Tennyson et al. (2004), Yurchenko et al. (2007), Makarewicz and Skalozub (2007), Mátyus et al. (2009a), Gatti and Iung (2009) and Chubb et al. (2020).

In particular, Carter and Handy (1986) developed a general polyatomic KEO; Eckart KEOs were explored by Szalay (2015b, 2017) and Rey (2019); KEOs of tetratomic molecules were reported in Gatti et al. (1998), Mladenovic (2000), Zhang et al. (1995), Xu et al. (2002); Yu and Muckerman (2002), Gatti and Nauts (2003) (Jacobi), Carter and Handy (1984); Bramley et al. (1991), Carter et al. (1997), Protasevich and Nikitin (2022) (valence), Schwenke (2003) (polyspherical), Bramley and Carrington (1993) and Mladenovic (2002) (Jacobi, polyspherical and Radau); other KEOs using polyspherical, Jacobi or Radau coordinates were developed in Aquilanti and Cavalli (1986), Chapuisat and Iung (1992), Wei and Carrington (1997), Gatti et al. (1998), Mladenovic (2000), Gatti et al. (2001), Leforestier et al. (2001), Yu (2002), Gatti and Nauts (2003), Makarewicz and Skalozub (2007), Gatti and Iung (2009) and Yu (2016).

The transformation from the Cartesian LF frame coordinate to the generalised body-frame system via the $\boldsymbol{t}$–$\boldsymbol{s}$ formalism is according to Sørensen (1979) and Watson (2004); the construction of KEO as inverse of the $\boldsymbol{g}$ kinetic tensor is from Fabri et al. (2011). Normal-mode KEO is from Watson (1968).

The general approach for the Taylor-type KEO is from Yurchenko et al. (2005a, 2007).

For the $\boldsymbol{t}$–$\boldsymbol{s}$-vector formalism, see Lukka (1995), Sørensen (1979), Watson (2004), Yurchenko et al. (2005a), Makarewicz and Skalozub (2007), Yurchenko et al. (2007) and Mátyus et al. (2009a).

Proof that it is not possible to fully eliminate the Coriolis coupling from the quantum-mechanical KEO can be found in Sørensen (1979). PAS KEO is from Meyer and Günthard (1968) and Sørensen (1979). Singularities in PAS KEO with are from Sørensen (1979).

KEO for a pyramidal and linear tetratomic molecule was derived by Schwenke (1996).

Some modern nuclear motion variational codes based on the body-frame KEOs are MORBID (Jensen, 1988), DVR3D (Tennyson et al., 2004), MULTIMODE (Carter et al., 2009), TROVE (Yurchenko et al., 2007), GENIUSH (Mátyus et al., 2009b; Fabri et al., 2011; Fábri et al., 2014), TENSOR packages (Rey et al.,

2010, 2012; Rey, 2019), DEWE (Mátyus et al., 2007), WAVR4 (Kozin et al., 2003, 2004), TetraVib (Yu and Muckerman, 2002), Schwenke's tetratomic code (Schwenke, 1996), etc.

MATERIALS NOT USED BUT WORTH MENTIONING

Development of KEOs using the Podolsky trick (Podolsky, 1928).

Automatic derivatives by Yachmenev and Yurchenko (2015); time-dependent MCTDH approaches (Manthe, 2008; Beck et al., 2000; Handy and Carter, 2004; Carter et al., 2009; Worth, 2020) including applications based on the Eckart frame (Sadri et al., 2014); collocation methods (Yang and Peet, 1988; Avila and Carrington, 2015; Carrington, 2021); Smolyak non-product quadrature grids (Avila and Carrington, 2009; Lauvergnat et al., 2016; Brown and Carrington, 2016) as well as nested quadrature grids (Thomas and Carrington, 2015).

Gauge formalism in separation of rotations and vibrational motions when constructing KEOs (Littlejohn and Reinsch, 1997).

KEO for a A_2B_2-type linear molecule using the $3N-5$ formalism with the curvilinear coordinates (Protasevich and Nikitin, 2022) and normal-mode coordinates (Schröder, 2022).

Hybrid approaches combining advantages of the variational and perturbational treatments include Fabri et al. (2014); Rey (2022).

See also the vibrational configuration interaction (VCI) (Neff and Rauhut, 2009) and rotational VCI (Erfort et al., 2020) techniques.

CHAPTER 4

Kinetic energy operator: Triatomic molecules

This chapter provides examples of construction of kinetic energy operators (KEOs) for triatomic molecules. Triatomic molecules are small enough to allow the corresponding KEOs in an analytic form and thus can serve to illustrate the generalised algorithms introduced above. We continue using the Sørensen procedure for construction of KEOs. The singularities appearing at linear geometries and possible solutions are discussed.

4.1 KEO OF XY_2 IN THE VALENCE COORDINATE REPRESENTATION

Let us now illustrate a derivation of KEO for a specific system of triatomic molecules using XY_2 as an example (e.g. water, H_2O). Following the methodology described above, the $3N = 9$ Cartesian coordinates $\boldsymbol{R}_n$ of the three nuclei associated with the LF frame need to be transformed to a set of three translational $\boldsymbol{R}^{\mathrm{CM}}$, three rotational θ, ϕ and χ and three vibrational coordinates ξ_1, ξ_2 and ξ_3 associated with the molecular-fixed frame as in Eq. (2.4). An example of a body frame for XY_2 at some arbitrary orientation in space and instantaneous configuration of the nuclei is shown in Fig. 4.1, which is centred at the nuclear centre-of-mass with the orientation of the molecule measured using the Euler angles θ, ϕ and χ. For simplicity, let us put aside any singularity problems associated with linear configurations by assuming that they are inaccessible or we know how to deal with them (see below). For the vibrational degrees of freedom, the valence and linearised coordinates will be considered as illustrations.

There is no unique choice for the orientation of the body-frame system. Among the typical criteria to set up the xyz axes are the following: (i) to simplify the coordinate transformation (and KEO), (ii) to minimise the coupling between the rotational and vibrational degrees of freedom and (iii) to help with the integration of the corresponding Schrödinger equation. Therefore to a large extent, the choice of the molecular frame is influenced by the choice of the internal (vibrational)

DOI: 10.1201/9780429154348-4

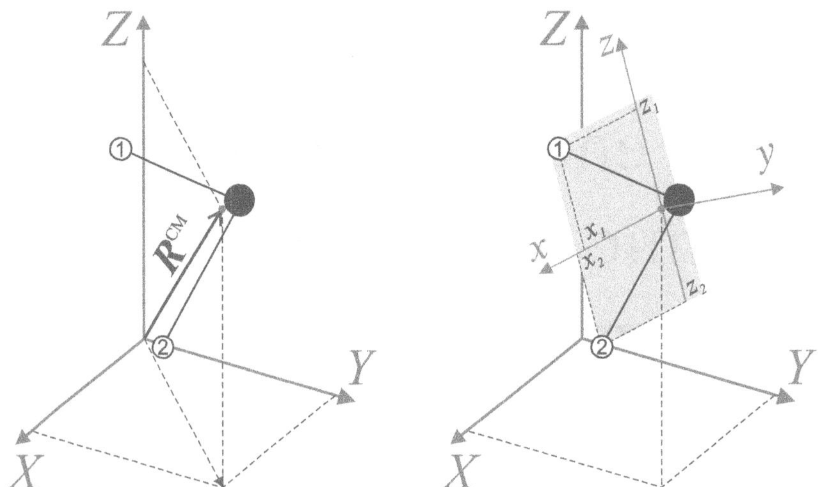

Figure 4.1 An illustration of laboratory- and body-fixed frames for an XY_2 molecule with a heavier X atom. The x axis is chosen parallel to the bisector of the inter-bond angle α.

coordinates as well as by the integration methods used, types of the basis sets employed, etc.

Let us first consider the valence-type coordinates as vibrational degrees of freedom for XY_2 as shown in Fig. 4.2 with α as an inter-bond angle and r_1 and r_2 as bond lengths. In order to define the orientation of the body-fixed orthogonal axes system xyz for an arbitrary instantaneous configuration of nuclei, three conditions are required. One way of defining these conditions is by explicitly specifying the body-instantaneous frame coordinates $\boldsymbol{r}_n^{\mathrm{BF}}$ as functions of the corresponding instantaneous vibrational degrees of freedom $\boldsymbol{\xi} = \{r_1, r_2, \alpha\}$, i.e. $\boldsymbol{r}_n^{\mathrm{BF}} = \boldsymbol{r}_n^{\mathrm{BF}}(\boldsymbol{\xi})$.

For a triatomic planar molecule XY_2, it makes sense to put two axes (e.g. x and z) in the molecular plane with the third axis (y) perpendicular to it, i.e. $y_1 = y_2 = y_3 = 0$, where the indices 1, 2 and 3 are used to label the atoms Y_1, Y_2 and X.

The in-plane coordinates should satisfy the centre-of-mass conditions:

$$\begin{aligned} x_1 m_1 + x_2 m_2 + x_3 m_3 &= 0, \\ z_1 m_1 + z_2 m_2 + z_3 m_3 &= 0. \end{aligned}$$

There are many (in principle infinite) ways to orient the in-plane axes x and z. For a system with (two) equivalent atoms, it also makes sense not to give any privileges to one of them when defining the internal axes. The bisector frame, where one of the axes (e.g. x) is placed parallel to the bisector of the valence angle Y_1–X–Y_2 (see Fig. 4.2), satisfies this requirement. Remember that the coordinate centre is placed at the centre-of-mass, not at the central X atom 3, and therefore it only coincides with the bisector of Y_1–X–Y_2 passing through the X nucleus for $r_1 = r_2$.

The z axis is another in-plane axis, which, together with the y axis, is chosen in the right-handed sense. A neat alternative to the valence bisector embedding is the Jacobi (bisector) frame as shown in Fig. 2.11 with the x axis making an angle $\alpha/2$ with any of the Jacobi vectors CM-Y_1 or CM–Y_2. Only at equilibrium do these embeddings correspond to a PAS (i.e. with a diagonal inertia matrix), where they also coincide.

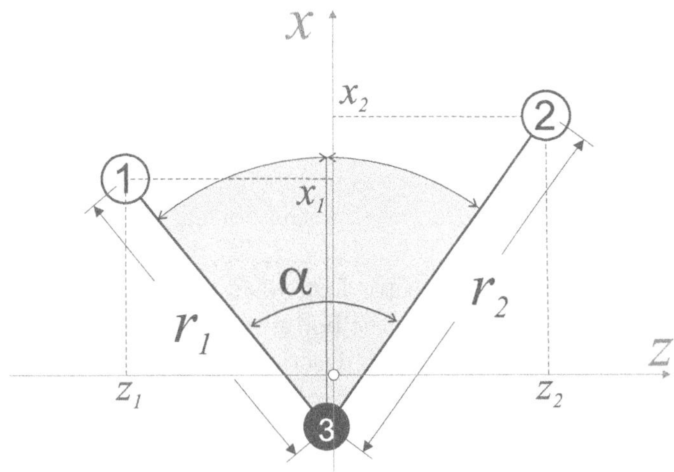

Figure 4.2 Body-fixed coordinates of an XY_2 molecule in the bisector frame. The CM is indicated by a white circle, and the y axis is orthogonal to the molecular frame in the right-handed sense. The nuclei are numbered as $1 = Y_1$, $2 = Y_2$ and $3 = X$. Atom 3 (X) is assumed to be heavier than atoms Y_1 and Y_2 here.

When defining the positions of $\boldsymbol{r}_n^{\mathrm{BF}} = \{x_1, x_2, x_3, z_1, z_2, z_3\}$ for arbitrary values of r_1, r_2 and α, it is instructive to initially place one of the atoms at the coordinate centre and then apply a uniform shift to the centre-of-mass to all coordinates. Then, for the body-fixed coordinates of nucleus 1, nucleus 2 and nucleus 3, we obtain

$$x_1 = r_1 \cos(\alpha/2) - \frac{m_Y \cos(\alpha/2)(r_1 + r_2)}{M_{\mathrm{tot}}}, \tag{4.1}$$

$$x_2 = r_2 \cos(\alpha/2) - \frac{m_Y \cos(\alpha/2)(r_1 + r_2)}{M_{\mathrm{tot}}}, \tag{4.2}$$

$$x_3 = 0 - \frac{m_Y \cos(\alpha/2)(r_1 + r_2)}{M_{\mathrm{tot}}}, \tag{4.3}$$

$$z_1 = -r_1 \sin(\alpha/2) + \frac{m_Y \sin(\alpha/2)(r_1 - r_2)}{M_{\mathrm{tot}}}, \tag{4.4}$$

$$z_2 = r_2 \sin(\alpha/2) + \frac{m_Y \sin(\alpha/2)(r_1 - r_2)}{M_{\mathrm{tot}}}, \tag{4.5}$$

$$z_3 = 0 + \frac{m_Y \sin(\alpha/2)(r_1 - r_2)}{M_{\mathrm{tot}}}, \tag{4.6}$$

which has the following form:

$$\boldsymbol{r}_n = \boldsymbol{r}'_n - \frac{\sum_{n=1}^{N} \boldsymbol{a}_n m_n}{\sum_{n=1}^{N} m_n}. \tag{4.7}$$

In Eqs. (4.1–4.6), M_{tot} as the total nuclear mass

$$M_{\text{tot}} \equiv \sum_{n=1}^{N} m_n = m_{\text{X}} + 2m_{\text{Y}}.$$

The valence coordinates r_1, r_2 and α are not the only coordinate choice for the XY_2 systems. Other popular alternatives include Jacobi coordinates (see Fig. 2.11), Radau coordinates (see Fig. 4.3) as well as the normal modes and linearised coordinates described above.

Let us now explore the $\boldsymbol{t}-\boldsymbol{s}$ vector formalism (see Section 3.2.1) to build the KEO for XY_2. Applying Eq. (3.26) to the body-fixed coordinates given in Eqs. (4.1–4.6), it is trivial to obtain an analytic description of the transformational 9×9 $\boldsymbol{t}$-matrix, especially with analytic algebra software.

The translational 9×3 part of $\boldsymbol{t}$ (with elements $t_{n\alpha,A}$) reads

$$t^{\text{trans}} = \left[\begin{array}{ccc} X & Y & Z \\ \hline 1 & 0 & 0 \\ 0 & 1 & 0 \\ 0 & 0 & 1 \\ 1 & 0 & 0 \\ 0 & 1 & 0 \\ 0 & 0 & 1 \\ 1 & 0 & 0 \\ 0 & 1 & 0 \\ 0 & 0 & 1 \end{array}\right] \begin{array}{c} n,\alpha \\ \hline \leftarrow 1,x \\ \leftarrow 1,y \\ \leftarrow 1,z \\ \leftarrow 2,x \\ \leftarrow 2,y \\ \leftarrow 2,z \\ \leftarrow 3,x \\ \leftarrow 3,y \\ \leftarrow 3,z \end{array} \tag{4.8}$$

where the last column with arrows is given to indicate the structure of the 9D vector ($n = 1, 2, 3$ and $\alpha = x, y, z$).

The rotational $\boldsymbol{t}_{n,g}$ (9×3) and vibrational parts $\boldsymbol{t}_{n,\lambda}$ (9×3) are given by

$$t^{\text{rot}} = \frac{1}{M} \times$$
$$\begin{bmatrix} 0 & -\sin\beta\,[m_{\text{X}}r_1 + m(r_1+r_2)] & 0 \\ \sin\beta\,[m_{\text{X}}r_1 + m(r_1+r_2)] & 0 & \cos\beta\,[m_{\text{X}}r_1 + m(r_1-r_2)] \\ 0 & -\cos\beta\,[m_{\text{X}}r_1 + m(r_1-r_2)] & 0 \\ 0 & \sin\beta\,[m_{\text{X}}r_2 + m(r_1+r_2)] & 0 \\ -\sin\beta\,[m_{\text{X}}r_2 + m(r_1+r_2)] & 0 & \cos\beta\,[m_{\text{X}}r_2 - m(r_1-r_2)] \\ 0 & -\cos\beta\,[m_{\text{X}}r_2 + m(r_1-r_2)] & 0 \\ 0 & m\sin\beta(r_1-r_2) & 0 \\ -m\sin\beta(r_1-r_2) & 0 & -m\cos\beta(r_1+r_2) \\ 0 & m\cos\beta(r_1+r_2) & 0 \end{bmatrix} \tag{4.9}$$

Table 4.1 The $\boldsymbol{s}$-matrix for the bisector valence coordinates frame of XY_2. The first column $(X, Y, Z, x, y, z, 1, 2, 3)$ and first row (n, α) indicate the correlation with the column and row indices, respectively. Here $\beta = \alpha/2$.

	$1x$	$1y$	$1z$	$2x$	$2y$	$2z$	$3x$	$3y$	$3z$
X	$\frac{m}{M}$	0	0	$\frac{m}{M}$	0	0	$\frac{m_X}{M}$	0	0
Y	0	$\frac{m}{M}$	0	0	$\frac{m}{M}$	0	0	$\frac{m_X}{M}$	0
Z	0	0	$\frac{m}{M}$	0	0	$\frac{m}{M}$	0	0	$\frac{m_X}{M}$
x	0	$\frac{1}{2r_1 \sin\beta}$	0	0	$\frac{-1}{2r_2 \sin\beta}$	0	0	$\frac{r_1 - r_2}{2r_1 r_2 \sin\beta}$	0
y	$\frac{-\sin\beta}{2r_1}$	0	$\frac{-\cos\beta}{2r_1}$	$\frac{\sin\beta}{2r_2}$	0	$\frac{-\cos\beta}{2r_2}$	$\frac{(r_2 - r_1)\sin\beta}{2r_1 r_2}$	0	$\frac{(r_1 + r_2)\cos\beta}{2r_1 r_2}$
z	0	$\frac{1}{2r_1 \cos\beta}$	0	0	$\frac{1}{2r_2 \cos\beta}$	0	0	$\frac{-(r_1 + r_2)}{2r_1 r_2 \cos\beta}$	0
1	$\cos\beta$	0	$-\sin\beta$	0	0	0	$-\cos\beta$	0	$\sin\beta$
2	0	0	0	$\cos\beta$	0	$\sin\beta$	$-\cos\beta$	0	$-\sin\beta$
3	$\frac{-\sin\beta}{2r_1}$	0	$\frac{-\cos\beta}{2r_1}$	$\frac{-\sin\beta}{2r_2}$	0	$\frac{\cos\beta}{2r_2}$	$\frac{\sin\beta(r_1 + r_2)}{2r_1 r_2}$	0	$\frac{(r_2 - r_1)\cos\beta}{2r_1 r_1}$

and

$$\boldsymbol{t}^{\text{vib}} = \frac{1}{M} \times \begin{bmatrix} \cos\beta(m_X + m) & -m\cos\beta & -\frac{1}{2}\sin\beta\,[r_1 m_X + m(r_1 - r_2)] \\ 0 & 0 & 0 \\ -\sin\beta(m_X + m) & -m\sin\beta & -\frac{1}{2}\cos\beta\,[r_1 m_X + m(r_1 + r_2)] \\ -m\cos\beta & \cos\beta(m_X + m) & \frac{1}{2}\sin\beta\,[-r_2 m_X + m(r_1 - r_2)] \\ 0 & 0 & 0 \\ m\sin\beta & \sin\beta(m_X + m) & \frac{1}{2}\cos\beta\,[r_2 m_X + m(r_1 + r_2)] \\ -m\cos\beta & -m\cos\beta & \frac{1}{2}m\sin\beta(r_1 + r_2) \\ 0 & 0 & 0 \\ m\sin\beta & -m\sin\beta & \frac{1}{2}m\cos\beta(r_1 - r_2) \end{bmatrix}$$

where $\beta = \alpha/2$ and $m = m_Y$. The determinant of the transformational $\boldsymbol{t}$-matrix,

$$\det(\boldsymbol{t}) = -r_1^2 r_2^2 \sin\alpha, \tag{4.10}$$

tends to zero at linearity ($\alpha = \pi$) and therefore will lead to a singularity in the KEO.

Having the 9×9 $\boldsymbol{t}$-matrix ($\{\boldsymbol{t}^{\text{trans}}, \boldsymbol{t}^{\text{rot}}, \boldsymbol{t}^{\text{vib}}\}$) defined, the 9×9 $\boldsymbol{s}$-matrix is obtained by solving Eq. (3.11) via

$$\boldsymbol{t} = \boldsymbol{s}^{-1}$$

with the solution shown in Table 4.1.

We now plug $\boldsymbol{s}_{\lambda,n}$ into Eq. (3.14) and obtain the kinetic energy factors $\boldsymbol{G}$ (so-called reciprocal reduced masses) as follows.

1. Vibrational part

$$G^{\rm vib}_{1,1} = G^{\rm vib}_{2,2} = \frac{1}{\mu_{XY}}, \tag{4.11}$$
$$G^{\rm vib}_{1,2} = G^{\rm vib}_{2,1} = \frac{\cos\alpha}{m_X}, \tag{4.12}$$
$$G^{\rm vib}_{1,3} = G^{\rm vib}_{3,1} = -\frac{\sin\alpha}{r_2 m_X}, \tag{4.13}$$
$$G^{\rm vib}_{2,3} = G^{\rm vib}_{3,2} = -\frac{\sin\alpha}{r_1 m_X}, \tag{4.14}$$
$$G^{\rm vib}_{3,3} = \frac{1}{\mu_{XY}}\left(\frac{1}{r_1^2}+\frac{1}{r_2^2}\right) - \frac{2\cos\alpha}{r_1 r_2 m_X}; \tag{4.15}$$

2. Pseudo-potential function

$$U(r_1, r_2, \alpha) = -\frac{\hbar^2}{8\sin^2(\alpha)}\left\{\frac{1+\sin^2(\alpha)}{\mu_{XY}}\left[\frac{1}{r_1^2}+\frac{1}{r_2^2}\right] + \frac{2\cos^3(\alpha)}{r_1 r_2 m_X}\right\}; \tag{4.16}$$

3. Coriolis part (with only non-zero elements shown):

$$G^{\rm Cor}_{1,y} = \frac{\sin\alpha}{2 r_2 m_X}, \tag{4.17}$$
$$G^{\rm Cor}_{2,y} = \frac{\sin\alpha}{2 r_1 m_X}, \tag{4.18}$$
$$G^{\rm Cor}_{3,y} = -\frac{1}{2\mu_{XY}}\left(\frac{1}{r_1^2}-\frac{1}{r_2^2}\right); \tag{4.19}$$

4. Rotational part (only non-zero elements):

$$G^{\rm rot}_{x,x} = \frac{1}{4\sin^2(\alpha/2)}\left[\frac{1}{\mu_{XY}}\left(\frac{1}{r_1^2}+\frac{1}{r_2^2}\right) - 2\frac{1}{m_X r_1 r_2}\right], \tag{4.20}$$
$$G^{\rm rot}_{x,z} = G^{\rm rot}_{z,x} = \frac{1}{2\mu_{XY}\sin\alpha}\left(\frac{1}{r_1^2}-\frac{1}{r_2^2}\right), \tag{4.21}$$
$$G^{\rm rot}_{y,y} = \frac{1}{4}\left[\frac{1}{\mu_{XY}}\left(\frac{1}{r_1^2}+\frac{1}{r_2^2}\right) + 2\frac{\cos\alpha}{m_X r_1 r_2}\right], \tag{4.22}$$
$$G^{\rm rot}_{z,z} = \frac{1}{4\cos^2(\alpha/2)}\left[\frac{1}{\mu_{XY}}\left(\frac{1}{r_1^2}+\frac{1}{r_2^2}\right) + \frac{2}{m_X r_1 r_2}\right]. \tag{4.23}$$

In Eqs. (4.11–4.23), μ_{XY} is the reduced mass given by

$$\frac{1}{\mu_{XY}} = \frac{1}{m_X} + \frac{1}{m_Y}$$

and we numbered the three coordinates r_1, r_2 and α as 1, 2 and 3, respectively.

Here we assume a non-Euclidian normalisation and the Wilson integration volume

$$dV = \sin\theta \, d\xi_1 d\xi_2 \dots d\xi_M \, d\phi \, d\theta \, d\chi$$

with ϕ, θ and χ as Euler angles.

An important observation of the valence coordinates form of $\boldsymbol{G}$ and U of XY_2 is that they can be represented as a sum-of-products of independent 1D terms, which makes them compatible with integrations using the product types of the basis sets. This is very useful for computing multi-dimensional matrix elements. For example, for vibrational 3D basis functions in a product form

$$\phi_{v_1,v_2,v_3}(r_1, r_2, \alpha) = \phi_{v_1}(r_1)\phi_{v_2}(r_2)\phi_{v_3}(\alpha)$$

and assuming a sum-of-products form for the pseudo-potential function

$$U = \sum_{l,m,n} u_{l,m,n} f_l(r_1) f_m(r_2) f_n(\alpha),$$

the matrix elements are conveniently decomposed into a sum-of-products of 1D integrals:

$$\begin{aligned} &\langle v_1, v_2, v_3 | U(r_1, r_2, \alpha) | v_1', v_2', v_3' \rangle \\ &\quad = \sum_{l,m,n} u_{l,m,n} \langle v_1 | f_l(r_1) | v_1' \rangle \langle v_2 | f_m(r_2) | v_2' \rangle \langle v_3 | f_n(\alpha) | v_3' \rangle, \end{aligned}$$

where the standard bra-ket notation was used for the matrix integral over the volume V

$$\langle v | f(\boldsymbol{\xi}) | v' \rangle = \int_V \phi_v^* f(\boldsymbol{\xi}) \phi_{v'} dV.$$

It is relatively straightforward to derive the KEO $\hat{T}$ analytically for an XY_2 molecule for the case of valence coordinates as in this example. For larger systems, the KEO derivations quickly become very complex even for analytic algebra programs. It is the inversion of the $\boldsymbol{t}$-matrix or the $\boldsymbol{J}$-matrix that makes the analytic approach so complicated. Another complication arises from the first and second derivatives of the elements of $\boldsymbol{s}$ that appear in the formulation of the pseudo-potential function U in Eq. (3.15). This is exactly the complexity that makes the numerical-based approaches discussed in Section 3.5.1 especially attractive.

4.1.1 Eckart frame

Let us now switch to the illustration of the Eckart embedding for XY_2 within the same $\boldsymbol{t}$–$\boldsymbol{s}$ approach. The Eckart conditions defining the body-fixed Cartesian coordinate system are fully represented by the three equations in Eq. (2.19). In principle these conditions, being linear in the Cartesian coordinates $\boldsymbol{r}_n$, are better suited for rectilinear coordinates (as will also be explored below). However, the (planar) triatomic molecules are simple enough for the Eckart frame to be defined also in conjunction with the curvilinear coordinates. The bisector frame defined above can be used as a reference frame, thus formulating the solution for the Eckart embedding as a simple rotation between them. In fact it is common to utilise reference frames of some convenient choice and then find a unitary transformation matrix to satisfy the three orientational constraints. For a general polyatomic molecule, the Eckart conditions can lead to transcendental equations on the orientation with no analytic solutions and are therefore better suited for numerical implementations.

A planar triatomic molecule is a special case for which the orientational conditions are reduced to a one-angle, in-plane rotation, which can be solved analytically. For the transformation from the reference bisector XY_2 frame ('bis') defined in Eq. (4.7) to the Eckart frame ('Eck')

$$\boldsymbol{r}_n^{\rm Eck} = \boldsymbol{T}(\phi)\boldsymbol{r}_n^{\rm bis}, \tag{4.24}$$

the 3×3 unitary transformation matrix is given by

$$\boldsymbol{T} = \begin{pmatrix} \cos\phi & 0 & -\sin\phi \\ 0 & 1 & 0 \\ \sin\phi & 0 & \cos\phi \end{pmatrix} \tag{4.25}$$

with ϕ as a single unknown parameter.

The equilibrium configuration of the body-fixed frame $\boldsymbol{r}_n^{\rm e}$ required by Eckart conditions is commonly chosen as a PAS with the coordinate centre at CM. For XY_2 shown in Fig. 4.2, the equilibrium Cartesian coordinates are hence given by

$$\begin{aligned} r_{1,x}^{\rm e} &= 2r_{\rm e}\cos\beta_{\rm e}\frac{m_{\rm X}}{M_{\rm tot}}, & r_{1,z}^{\rm e} &= -r_{\rm e}\sin\beta_{\rm e}, \\ r_{2,x}^{\rm e} &= 2r_{\rm e}\cos\beta_{\rm e}\frac{m_{\rm X}}{M_{\rm tot}}, & r_{2,z}^{\rm e} &= r_{\rm e}\sin\beta_{\rm e}, \\ r_{3,x}^{\rm e} &= -2r_{\rm e}\cos\beta_{\rm e}\frac{m_{\rm Y}}{M_{\rm tot}}, & r_{3,z}^{\rm e} &= 0, \end{aligned} \tag{4.26}$$

where $M_{\rm tot} = m_{\rm X} + 2m_{\rm Y}$ is the total mass of the nuclei and $\beta_{\rm e} = \alpha_{\rm e}/2$ and the nuclei are numbered as $1 = {\rm Y}_1$, $2 = {\rm Y}_2$ and $3 = {\rm X}$.

Using the bisector-frame instantaneous positions $\boldsymbol{r}_n^{\rm bis}$ from Eqs. (4.1–4.6) as a reference, after the rotation about the y axis through ϕ radians in Eq. (4.24), for the new (Eckart) frame, we obtain

$$
\begin{aligned}
r_{1,x}^{\mathrm{Eck}} &= \left[r_1 \cos(\beta - \phi)(m_{\mathrm{X}} + m_{\mathrm{Y}}) - r_2 \cos(\beta + \phi) m_{\mathrm{Y}}\right] \frac{1}{M_{\mathrm{tot}}}, \\
r_{1,z}^{\mathrm{Eck}} &= \left[-r_1 \sin(\beta - \phi)(m_{\mathrm{X}} + m_{\mathrm{Y}}) - r_2 \sin(\beta + \phi) m_{\mathrm{Y}}\right] \frac{1}{M_{\mathrm{tot}}}, \\
r_{2,x}^{\mathrm{Eck}} &= \left[-r_1 \cos(\beta - \phi) m_{\mathrm{Y}} + r_2 \cos(\beta + \phi)(m_{\mathrm{X}} + m_{\mathrm{Y}})\right] \frac{1}{M_{\mathrm{tot}}}, \\
r_{2,z}^{\mathrm{Eck}} &= \left[r_1 \sin(\beta - \phi) m_{\mathrm{Y}} + r_2 \sin(\beta + \phi)(m_{\mathrm{X}} + m_{\mathrm{Y}})\right] \frac{1}{M_{\mathrm{tot}}}, \\
r_{3,x}^{\mathrm{Eck}} &= \left[-r_1 \cos(\beta - \phi) - r_2 \cos(\beta + \phi)\right] \frac{m_{\mathrm{Y}}}{M_{\mathrm{tot}}}, \\
r_{3,z}^{\mathrm{Eck}} &= \left[r_1 \sin(\beta - \phi) - r_2 \sin(\beta + \phi)\right] \frac{m_{\mathrm{Y}}}{M_{\mathrm{tot}}}.
\end{aligned}
\tag{4.27}
$$

Applying the Eckart conditions in Eq. (2.19) to $\boldsymbol{r}_n^{\mathrm{Eck}}$ and $\boldsymbol{r}_n^{\mathrm{e}}$ above, we find that the equations for the x and y components are fulfilled automatically, while the Eckart equation for y becomes

$$
r_{\mathrm{e}}\, m_{\mathrm{Y}} \sin(\alpha_{\mathrm{e}}/2) \left[r_1 \cos(\beta - \phi) - r_2 \cos(\beta + \phi)\right] = 0, \tag{4.28}
$$

which leads to the simple solution for ϕ as

$$
\phi = \arctan\left[\cot(\beta) \frac{r_2 - r_1}{r_1 + r_2}\right].
$$

The $\boldsymbol{r}_n^{\mathrm{Eck}}$ in Eq. (4.27) can now be used in Eq. (3.26) in the same manner as for the bisector frame above to obtain $\boldsymbol{t}$, $\boldsymbol{s}$, $\boldsymbol{G}$ and U, which, however, will lead to more complicated expressions for $\boldsymbol{s}$ and therefore for $\boldsymbol{G}$ and U. It should be already clear, that for larger molecules, a purely analytic description of the KEO can become quickly unsustainable, which strengthens the advantage of the numerical, automatic formulation of the KEO construction.

It should also be noted that the Eckart system is considered to be efficient for separating the vibrational and rotational motion and reducing the Coriolis interaction, which, e.g., can help with the convergence of variational solutions. The disadvantage of the Eckart conditions is that, because of the formulation in terms of Cartesian displacements, it is more difficult to use for curvilinear coordinates. The geometrically defined frames such as the bisector, bond, Radau, atom-diatom, etc., are therefore a more popular choice when curvilinear coordinates are used. Any negative impacts from the larger Coriolis coupling ensued are then mitigated by employing larger vibrational basis sets, if feasible.

4.1.2 Sørensen's solution

As we showed above (see Section 3.4), with the help of the Sørensen approach, the inversion of a general $3N \times 3N$ $\boldsymbol{t}$-matrix can be simplified to an inversion of a 3×3

matrix $\boldsymbol{J}$. The approach is based on the formulation of the three $\boldsymbol{C}$ vectors as three rotational constraints in Eq. (3.36). For an XY_2 molecule in the xz plane, confining the three nuclei in the xz plane gives us two conditions

$$C_x \equiv r_{1y} - r_{3y} = 0,$$
$$C_y \equiv r_{2y} - r_{3y} = 0,$$

and the third condition is to place the x vector along the bisector of α using the following identity for $\tan\alpha/2$ in terms of the components of the nuclear coordinates

$$\tan\frac{\alpha}{2} = -\frac{r_{1z} - r_{3z}}{r_{1x} - r_{3x}} = \frac{r_{2z} - r_{3z}}{r_{2x} - r_{3x}}$$

or equivalently

$$-(r_{1z} - r_{3z})(r_{2x} - r_{3x}) = (r_{2z} - r_{3z})(r_{1x} - r_{3x}).$$

The three constraints C_g ($g = x, y, z$) to define an x-bisector frame are then given by

$$\begin{aligned} C_x &\equiv r_{1y} - r_{3y} = 0, \\ C_y &\equiv r_{2y} - r_{3y} = 0, \\ C_z &\equiv (r_{1z} - r_{3z})(r_{2x} - r_{3x}) + (r_{2z} - r_{3z})(r_{1x} - r_{3x}) = 0. \end{aligned}$$

These lead to the following non-zero components of $\boldsymbol{c}_{g,n}$ obtained as 1st derivatives of C_g with respect to $\boldsymbol{r}_n$:

$$\begin{aligned} c_{x,3,y} &= -1, \quad c_{x,1,y} = 1, \\ c_{y,3,y} &= -1, \quad c_{y,2,y} = 1, \\ c_{z,3,x} &= -(r_{1z} - r_{3z}) - (r_{2z} - r_{3z}) = (r_1 - r_2)\sin\beta, \\ c_{z,3,z} &= -(r_{2x} - r_{3x}) - (r_{1x} - r_{3x}) = -(r_1 + r_2)\cos\beta, \\ c_{z,1,x} &= (r_{2z} - r_{3z}) = r_2\sin\beta, \\ c_{z,1,z} &= (r_{2x} - r_{3x}) = r_2\cos\beta, \\ c_{z,2,x} &= (r_{1z} - r_{3z}) = -r_1\sin\beta, \\ c_{z,2,z} &= (r_{1x} - r_{3x}) = r_1\cos\beta, \end{aligned}$$

where $\beta = \alpha/2$.

The 3×3 matrix $\boldsymbol{J}$ is obtained using Eq. (3.42) with the rotational $\boldsymbol{t}$-matrix elements of XY_2 from Eq. (4.9) and is given by

$$\boldsymbol{J} = \begin{pmatrix} r_1\sin(\beta) & 0 & r_1\cos(\beta) \\ -r_2\sin(\beta) & 0 & r_2\cos(\beta) \\ 0 & -2r_1r_2 & 0 \end{pmatrix}. \tag{4.29}$$

This matrix, which has the determinant

$$\det(\boldsymbol{J}) = 2r_1^2r_2\sin\alpha,$$

can be inverted either analytically or indeed numerically at any given configuration of $r_1, r_2, \beta = \alpha/2$ (except for $\beta = 0$, where the determinant is zero) to obtain the matrix $\boldsymbol{\eta}$:

$$\eta = \begin{pmatrix} \frac{1}{2r_1 \sin(\beta)} & -\frac{1}{2r_2 \sin(\beta)} & 0 \\ 0 & 0 & -\frac{1}{2r_1 r_2} \\ \frac{1}{2r_1 \cos(\beta)} & \frac{1}{2r_2 \cos(\beta)} & 0 \end{pmatrix}.$$

In turn, the latter gives the rotational $\boldsymbol{s}$-matrix via Eq. (3.41), which is obtained identical to that in Eq. (4.1) up to an arbitrary phase.

The vibrational elements of the $\boldsymbol{s}$-matrix, $\boldsymbol{s}_{i,n}^{(\mathrm{vib})}$, are obtained as derivatives of $\xi_1 = r_1, \xi_2 = r_2$ and $\xi_3 = \alpha$, with respect to the Cartesian coordinates; see Eq. (3.21):

$$s_{\lambda,n\alpha}^{(\mathrm{vib})} = \frac{\partial \xi_\lambda}{\partial r_{n\alpha}},$$

with $\lambda = 1, 2, 3$. The derivatives of the valence coordinates with respect to the Cartesian displacements are given in Eq. (2.42), where the Cartesian coordinates need to be expressed in terms of the valence coordinates r_1, r_2, α via the corresponding definition of the body-frame coordinates $\boldsymbol{r}_n^{\mathrm{BF}} = \boldsymbol{r}_n^{\mathrm{BF}}(r_1, r_2, \alpha)$, e.g. Eq. (4.7) for the bisector frame or Eq. (4.27) for the Eckart frame.

These manipulations lead to the expressions of $\boldsymbol{s}_{i,n}^{(\mathrm{vib})}$ identical to those derived above using the $\boldsymbol{t}$–$\boldsymbol{s}$ approach collected in Table 4.1.

4.1.3 PAS frame

Considering the importance of the PAS for molecular physics, it can be also considered as one of the choices of the molecular embedding. The PAS orientational conditions are defined as follows:

$$\sum_n m_n r_{n,\alpha} r_{n,\beta} = 0, \quad \alpha \neq \beta, \tag{4.30}$$

where $\alpha, \beta = x, y, z$. Let us now explore the PAS frame for the XY_2 valence system. We can follow the same approach used for the Eckart frame (see Eq. (4.24)) and obtain the PAS frame from the bisector frame as a reference via the rotation in Eq. (4.25), which can be still solved analytically. The angle ϕ to satisfy the PAS conditions in Eq. (4.30) is given by

$$\tan 2\phi = \frac{\sin\alpha(m_{\mathrm{X}} + m_{\mathrm{Y}})(r_1^2 - r_2^2)}{\cos\alpha(m_{\mathrm{X}} + m_{\mathrm{Y}})(r_1^2 + r_2^2) - 2m_{\mathrm{Y}} r_1 r_2}. \tag{4.31}$$

Using the Sørensen approach, the KEO for the PAS embedding can be built analytically. The required $\boldsymbol{C}$ constraints vectors for the PAS conditions in Eq. (4.25) are given by

$$C_g = \sum_{n\beta\gamma} \epsilon_{g\beta\gamma}^2 m_n r_{n\beta} r_{n\gamma} = 0 \tag{4.32}$$

with the elements of $\boldsymbol{c}_{g,n}$

$$c_{g,n\beta} = 2\sum_{\gamma} \epsilon^2_{g\beta\gamma} m_n r_{n\gamma}.$$

Using these vectors in Eq. (3.39) and then in Eqs. (3.41,3.42), we obtain

$$s_{g,n\beta} = \sum_{\gamma} \frac{\epsilon^2_{g\beta\gamma} m_n r_{n\gamma}}{(I_{\gamma\gamma} - I_{\beta\beta})},$$

where I_{xx}, I_{yy} and I_{zz} are diagonal elements of the inertia tensor, which is diagonal in the PAS by construction. The main drawback of the PAS embedding is that the transformation matrix $\boldsymbol{s}$ and therefore kinetic energy factors $\boldsymbol{G}$ and U may become large in the case of accidentally similar values of the instantaneous inertia moments, $I_{\gamma\gamma} \sim I_{\beta\beta}$.

4.1.4 Radau frame

We saw in Section 4.1 that the bisector frame with the valence coordinates offers a relatively simple form of KEO. Another useful frame often used for three-atomic molecules is the Radau embedding. It is associated with the so-called canonical point and associated 'orthogonal' coordinates, which are illustrated in Fig. 4.3. The canonical point B is defined using the following Radau's geometric condition (Radau, 1868)

$$\overline{\mathrm{BD}}^2 = \overline{\mathrm{OD}} \cdot \overline{\mathrm{CD}},$$

where D is the centre-of-mass of the nuclei 1 and 2, and $\overline{\mathrm{BD}}$ is a geometric mean of $\overline{\mathrm{OD}}$ and $\overline{\mathrm{CD}}$. As we will see, these coordinates lead to an even simpler form of KEO with the vibrational part being diagonal and the Coriolis part reduced to the only term $G^{\mathrm{Cor}}_{\rho,z}$. The Radau embedding is defined by the two lengths of vectors linking the canonical point B with the nuclei 1 and 2 (r_1 and r_2), and the angle between these two vectors.

The instantaneous configuration in the Radau frame is given by

$$\begin{aligned}
r_{1,x} &= r_1 \cos(\theta/2) - \mathrm{CM}_x, \\
r_{1,z} &= -r_1 \sin(\theta/2) - \mathrm{CM}_z, \\
r_{2,x} &= r_2 \cos(\theta/2) - \mathrm{CM}_x, \\
r_{2,z} &= r_2 \sin(\theta/2) - \mathrm{CM}_z, \\
r_{3,x} &= \frac{1}{2}(r_1 + r_2)\cos(\theta/2)\left(1 - \sqrt{\frac{M_{\mathrm{tot}}}{m_X}}\right) - \mathrm{CM}_x, \\
r_{3,z} &= \frac{1}{2}(r_2 - r_1)\sin(\theta/2)\left(1 - \sqrt{\frac{M_{\mathrm{tot}}}{m_X}}\right) - \mathrm{CM}_z,
\end{aligned}$$

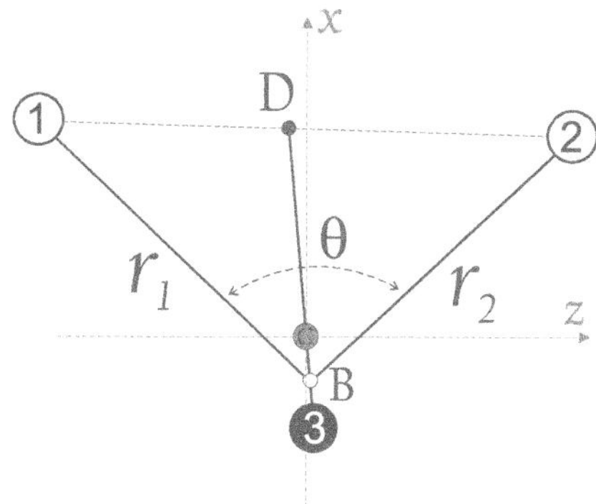

Figure 4.3 Radau frame and coordinates for an XY_2 molecule. D is the centre-of-mass of the atoms 1 and 2, C is the centre-of-mass and B is the canonical point defined as $\overline{\mathrm{BD}}^2 = \overline{\mathrm{OD}} \cdot \overline{\mathrm{CD}}$.

where $r_{n,y} = 0$ $(n = 1, 2, 3)$ and the centre-of-mass vector $\boldsymbol{C}$ is given by

$$C_x = \frac{1}{2}(r_1 + r_2)\cos(\theta/2)\left(1 - \sqrt{\frac{m_X}{M_{\mathrm{tot}}}}\right), \tag{4.33}$$

$$C_z = \frac{1}{2}(r_2 - r_1)\sin(\theta/2)\left(1 - \sqrt{\frac{m_X}{M_{\mathrm{tot}}}}\right). \tag{4.34}$$

As one would expect, the transformation to the Radau frame becomes singular at linear geometries, as can be seen from the determinant of the transformational matrix $\boldsymbol{t}$:

$$\det(\boldsymbol{t}) = -R_1^2 R^2 \sin(\theta)\left(\frac{M_{\mathrm{tot}}}{m_X}\right)^{3/2},$$

which tends to zero at $\theta \to \pi$ or 0. The KEO in the Radau representation has a very simple form indeed. The vibrational part is diagonal, as expected for the orthogonal coordinates:

$$G_{11}^{\mathrm{vib}} = \frac{1}{m_Y},$$

$$G_{22}^{\mathrm{vib}} = \frac{1}{m_Y},$$

$$G_{33}^{\mathrm{vib}} = \left(\frac{1}{m_Y r_1^2} + \frac{1}{m_Y r_2^2}\right),$$

while the Coriolis and rotational parts are also very compact:

$$G_{xx}^{\mathrm{rot}} = \frac{1}{4\,m_Y \sin^2(\theta/2)}\left[\frac{1}{r_1^2} + \frac{1}{r_2^2}\right],$$

$$G_{yy}^{\mathrm{rot}} = \frac{1}{4\,m_Y}\left[\frac{1}{m_Y r_1^2} + \frac{1}{m_Y r_2^2}\right],$$

$$G_{zz}^{\rm rot} = \frac{1}{4\,m_Y \cos^2(\theta/2)}\left[\frac{1}{r_1^2}+\frac{1}{r_2^2}\right],$$
$$G_{xz}^{\rm rot} = -\frac{1}{2\,m_Y\,\sin\theta}\left[\frac{1}{r_1^2}-\frac{1}{r_2^2}\right].$$

The only non-zero Coriolis term is

$$G_{3y}^{\rm Cor} = \frac{1}{2\,m_Y}\left[\frac{1}{r_1^2}-\frac{1}{r_2^2}\right]$$

and the pseudo-potential function is given by

$$U = \frac{\cos^2\theta - 2}{8\,\sin^2\theta}\left(\frac{1}{m_Y r_1^2}+\frac{1}{m_Y r_2^2}\right).$$

The singularities appear in U, $G_{xz}^{\rm rot}$ and $G_{zz}^{\rm rot}$ at $\theta = \pi$ (open linear configuration of the nuclei), while $G_{xz}^{\rm rot}$, $G_{xx}^{\rm rot}$ and U are also singular at $\theta = 0$ (closed linear configuration).

4.2 XYZ MOLECULE

For an asymmetric XYZ-type molecule, a geometrically motivated choice is to place the z axis parallel to one of the bonds, e.g. the heavier X–Y as given by

$$x_1 = 0 - X^{\rm CM}, \qquad z_1 = 0 - Z^{\rm CM},$$
$$x_2 = 0 - X^{\rm CM}, \qquad z_2 = r_1 - Z^{\rm CM},$$
$$x_3 = -r_2 \sin\rho - X^{\rm CM}, \qquad z_3 = -r_2 \cos\rho - Z^{\rm CM},$$

where r_1, r_2 and ρ are instantaneous values of the bond lengths and angle coordinates ($\alpha = \pi - \rho$ is the inter-bond angle) and the molecule is in the xz-plane; see Fig. 4.4. Here the centre-of-mass shift is given by

$$X^{\rm CM} = -\frac{m_Z\, r_2\, \sin\alpha}{m_X + m_Y + m_Z}, \tag{4.35}$$
$$Y^{\rm CM} = 0, \tag{4.36}$$
$$Z^{\rm CM} = \frac{m_Y\, r_1 + m_Z\, r_2\, \cos\alpha}{m_X + m_Y + m_Z}. \tag{4.37}$$

Applying the $\boldsymbol{t}$–$\boldsymbol{s}$ formalism, the transformation matrix elements $\boldsymbol{t}_{n,\lambda}$ are easily obtained using Eq. (3.26). The $\boldsymbol{t}$-matrix has the determinant

$$\det(\boldsymbol{t}) = r_1^2 r_2^2 \sin\rho,$$

which vanishes at linearity when $\rho = 0$.

Other choices leading to more efficient but also more cumbersome KEOs include the Eckart, PAS or non-rigid reference frames; see Eq. (2.19), Eq. (2.22) and Eq. (3.89), respectively.

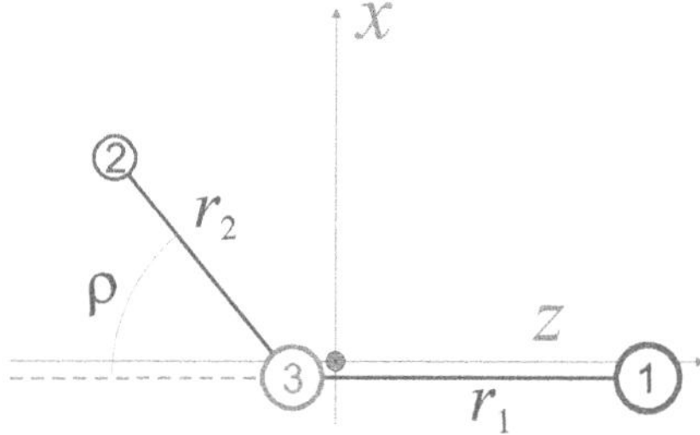

Figure 4.4 Body-fixed coordinates for an XYZ-type molecule. The xyz frame is centred at the nuclear centre-of-mass.

Assuming the internal coordinates $\xi_1 = r_1$, $\xi_2 = r_2$ and $\xi_3 = \rho$, the $\boldsymbol{s}$-matrix is obtained as an inverse of $\boldsymbol{t}$. The rotational elements are given by

$$
\begin{aligned}
s^{(\mathrm{rot})}_{x,1,y} &= -s^{(\mathrm{rot})}_{x,2,y} = -s^{(\mathrm{rot})}_{y,1,x} = s^{(\mathrm{rot})}_{y,2,x} = \frac{1}{r_1},\\
s^{(\mathrm{rot})}_{z,1,y} &= \frac{r_2\cos\rho + r_1}{r_1 r_2 \sin\rho}, \quad s^{(\mathrm{rot})}_{z,2,y} = -\frac{\cos\rho}{r_1\sin\rho},\\
s^{(\mathrm{rot})}_{z,3,y} &= \frac{\cos\rho}{r_2\sin\rho}
\end{aligned}
$$

and vibrational elements are given by

$$
\begin{aligned}
s^{(\mathrm{vib})}_{1,1,z} &= -s^{(\mathrm{vib})}_{1,2,z} = 1, \quad s^{(\mathrm{vib})}_{2,1,x} = -s^{(\mathrm{vib})}_{2,3,z} = \sin\rho,\\
s^{(\mathrm{vib})}_{2,1,z} &= -s^{(\mathrm{vib})}_{2,3,z} = \cos\rho, \quad s^{(\mathrm{vib})}_{3,1,x} = -\frac{r_2 + r_1\cos\rho}{r_1 r_2},\\
s^{(\mathrm{vib})}_{3,1,z} &= -s^{(\mathrm{vib})}_{3,3,z} = -\frac{\sin\rho}{r_2},\\
s^{(\mathrm{vib})}_{3,2,x} &= -\frac{1}{r_1}, \quad s^{(\mathrm{vib})}_{3,3,x} = -\frac{\cos\rho}{r_2},
\end{aligned}
$$

where only non-zero elements are shown. The rotational components with singularities at $\rho = 0$ are $s^{(\mathrm{rot})}_{z,2,y}$ and $s^{(\mathrm{rot})}_{z,2,y}$.

For the KEO, we obtain

$$
\begin{aligned}
G^{\mathrm{rot}}_{x,x} &= G^{\mathrm{rot}}_{y,y} = \frac{1}{\mu_{XY} r_1^2},\\
G^{\mathrm{rot}}_{x,z} &= G^{\mathrm{rot}}_{z,x} = \frac{\cos\rho}{\mu_{XY} r_1^2 \sin\rho} + \frac{1}{r_1 r_2 \sin\rho\, m_X},\\
G^{\mathrm{rot}}_{z,z} &= \frac{1}{\sin^2\rho}\left[\frac{\cos^2\rho}{\mu_{XY}\, r_1^2} + \frac{\cos^2\rho}{\mu_{XZ} r_2^2} + \frac{2\cos\rho}{m_X r_1 r_2}\right],
\end{aligned}
$$

$$G^{\rm vib}_{1,1} = \frac{1}{\mu_{XY}}, \quad G^{\rm vib}_{2,2} = \frac{1}{\mu_{XZ}},$$
$$G^{\rm vib}_{1,2} = G^{\rm vib}_{2,1} = -\frac{\cos\rho}{m_X}, \quad G^{\rm vib}_{1,3} = G_{3,1} = \frac{\sin\rho}{r_2 m_X},$$
$$G^{\rm vib}_{3,3} = \frac{1}{r_1^2\mu_{XY}} + \frac{1}{r_2^2\mu_{XZ}} + \frac{2\cos\rho}{r_1 r_2 m_X}$$

and

$$G^{\rm Cor}_{2,y} = -\frac{\sin\rho}{r_1\, m_X}, \quad G^{\rm Cor}_{3,y} = -\frac{1}{r_1^2\,\mu_{XY}} - \frac{\cos\rho}{r_1\, r_2\, m_X},$$

where the following notations of the reduced masses are used:

$$\mu_{XY} = \frac{1}{m_X} + \frac{1}{m_Y}, \quad \mu_{XZ} = \frac{1}{m_X} + \frac{1}{m_Z}.$$

Finally, the pseudo-potential function for this frame is given by

$$U = \frac{\hbar^2}{8\sin^2\rho}\left[\left(\frac{1}{\mu_{XY}r_1^2} + \frac{1}{\mu_{XZ}r_2^2}\right)(\cos^2\rho - 2) - \frac{2\cos^2\rho}{m_X r_1 r_2}\right].$$

Other popular coordinate choices for a non-symmetric XYZ molecule include the atom-diatom scattering coordinates shown in Fig. 4.5, where r is the bond length between the atoms 2 and 3 and R is the distance from atom 1 to the centre-of-mass of diatom 2–3 and the angle measured between the associated vectors $\mathbf{r}$ and $\mathbf{R}$. These are Jacobi-type coordinates. The molecular frame is chosen either along the bond $\mathbf{R}$ or along the bond $\mathbf{r}$.

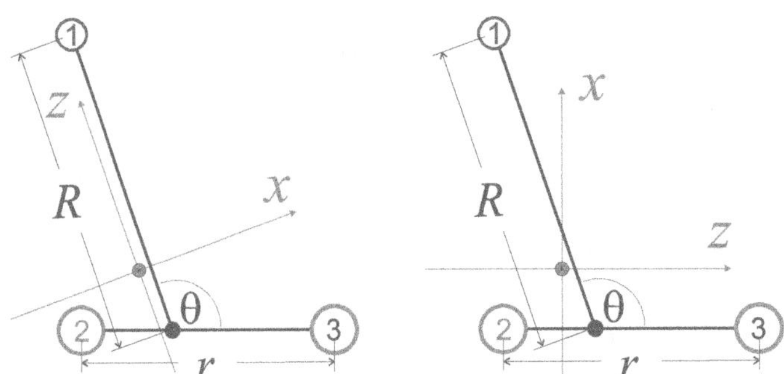

Figure 4.5 The atom-diatom scattering coordinates (Jacobi-type) used for an XYZ-type molecule, with the distance R measured from the diatom 2–3 centre-of-mass and atom 3. Two alternative xyz embeddings are shown with the z axis along the bond $\mathbf{R}$ and with the z axis along the bond $\mathbf{r}$.

4.3 LINEARISED COORDINATES FOR XY_2

Construction of the KEO in the linearised coordinates (Eckart frame) requires two main ingredients, the body-frame Cartesian coordinates at the equilibrium and their equilibrium derivatives with respect to $\boldsymbol{\xi}^{\ell}$, i.e. the $\boldsymbol{B}$-matrix; see Section 3.7.2. For the XY_2 system, the equilibrium structure $\boldsymbol{r}_n^{\rm e}$ satisfying the CM and PAS conditions is defined in Eq. (4.26). The derivatives of $\boldsymbol{r}_n$ taken at the equilibrium give the following matrix elements $B_{\lambda,n\alpha}$:

$$\begin{bmatrix} \boldsymbol{B}_{1,1} \\ \boldsymbol{B}_{1,2} \\ \boldsymbol{B}_{1,3} \end{bmatrix} = \begin{bmatrix} \sin\beta_{\rm e} & 0 & -\cos\beta_{\rm e} \\ 0 & 0 & 0 \\ -\sin\beta_{\rm e} & 0 & \cos\beta_{\rm e} \end{bmatrix}$$

$$\begin{bmatrix} \boldsymbol{B}_{2,1} \\ \boldsymbol{B}_{2,2} \\ \boldsymbol{B}_{2,3} \end{bmatrix} = \begin{bmatrix} 0 & 0 & 0 \\ \sin\beta_{\rm e} & 0 & -\cos\beta_{\rm e} \\ -\sin\beta_{\rm e} & 0 & -\cos\beta_{\rm e} \end{bmatrix}$$

$$\begin{bmatrix} \boldsymbol{B}_{3,1} \\ \boldsymbol{B}_{3,2} \\ \boldsymbol{B}_{3,3} \end{bmatrix} = \begin{bmatrix} -\frac{\cos\beta}{r_{\rm e}} & 0 & -\frac{\sin\beta}{r_{\rm e}} \\ -\frac{\cos\beta}{r_{\rm e}} & 0 & \frac{\sin\beta}{r_{\rm e}} \\ 2\frac{\cos\beta}{r_{\rm e}} & 0 & 0 \end{bmatrix},$$

where $\beta_{\rm e} = \alpha_{\rm e}/2$.

Once this property is defined, the KEO is trivially constructed by following the $\boldsymbol{t}$–$\boldsymbol{s}$ procedure as detailed, e.g., in Section 3.7.2.

4.4 SINGULARITIES IN KEO OF LINEAR AND QUASI-LINEAR TRIATOMIC MOLECULES

When the geometry of a polyatomic molecule becomes linear, the standard coordinate transformation involving the three Euler angles ϕ, θ and χ becomes singular due to degeneracy associated with the molecular orientation. In the case of XY_2, our standard choice of the $3N-6$ curvilinear coordinates includes r_1, r_2 and α (see Fig. 4.2) giving rise to a zero-valued transformation Jacobian at the linear configuration ($\alpha = 180°$), see Eq. (4.10), due to the Euler angle χ being undefined and thus leading to singular kinetic energy terms G_{zz}, U and $G_{x,z}$; see Eqs. (4.16, 4.21 and 4.23).

The resolution of this singularity at the linear configuration is usually associated with the following two methods. According to method 1, which is also referred to as the $(3N-5)$-method, (quasi-)linear molecules are treated using $3N-5$ coordinates with a transformation involving only two Eckart angles and four generalised vibrational modes. As we will illustrate below, all coordinates are well defined in this case and therefore no singularities appear.

According to method 2, or the $(3N-6)$ method, (quasi-)linear molecules are effectively treated as non-linear systems with ensuing singularity resolved by selecting appropriate basis set functions. We will show that the singularities of the type

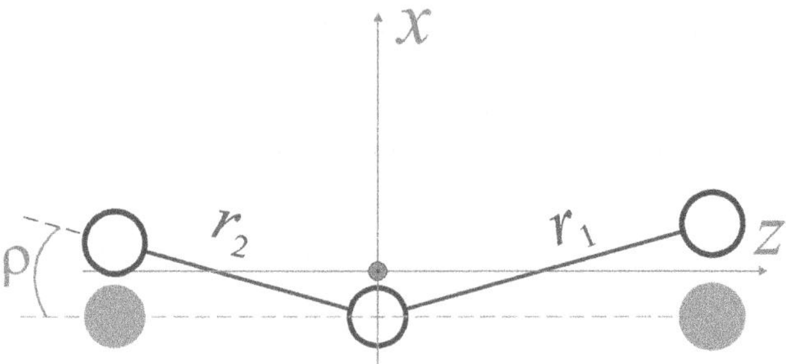

Figure 4.6 Body-fixed coordinates for an XY_2-type molecule with ρ measuring the bending angle from the linear configuration.

$1/\sin^2\rho$ and $1/\sin\rho$ in G_{zz}, U and $G_{x,z}$ can be cancelled by choosing the basis function with the appropriate counteractive factors ($\sim (\sin\rho)^{k+1/2}$ or $\sim \rho^{k+1/2}$, where k is the rotational quantum number). Here ρ is the bending coordinates $\rho = \alpha - 180°$ chosen to be zero at linear geometry; see Fig. 4.6.

4.4.1 Method 1: $3N - 5$ case or four rectilinear coordinates

Since the singularity issue in the KEO is associated with one of the Euler angles, χ, becoming undefined at linearity, the main point of the $3N - 5$ approach is to combine the corresponding degree of freedom with the vibrational ones and use internal coordinates that can be defined at any nuclear configuration.

Obviously, a trivial inclusion of the angle χ into a set of vibrational coordinates ξ_λ ($\lambda = 1 \ldots 3N - 5$) in Eq. (3.5) by combining it with the valence coordinates r_1, r_2 and α does not help. The nature of χ becomes undefined when α approaches zero, where these two angles define the orientation of the outer nuclei Y_1 and Y_2 (see Fig. 4.7) relative to the molecular z axis. A natural solution to this problem is to

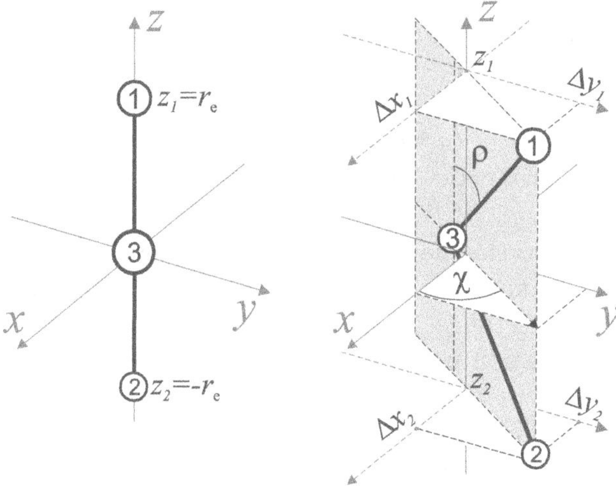

Figure 4.7 A $3N - 5$ rectilinear frame used for linear triatomic molecule.

use two rectilinear coordinates to describe the orientation of the molecule instead of the pair χ and α as follows.

A simple non-singular KEO can be constructed for four Cartesian-like coordinates as follows. Let us define the equilibrium configuration $r^{\rm e}_{i\alpha}$ of a linear molecule along the molecular z axis (centred at $z = 0$), as shown in Figure 4.7. In the molecular frame, we introduce the following four rectilinear coordinates to describe any instantaneous configuration of the nuclei:

$$
\begin{aligned}
\xi_1 \equiv& \Delta z_1 = z_1 - z_3, \\
\xi_2 \equiv& -\Delta z_2 = -(z_2 - z_3), \\
\xi_3 \equiv& \Delta x_1 + \Delta x_2 = (x_1 - x_3) + (x_2 - x_3), \\
\xi_4 \equiv& \Delta y_1 + \Delta y_2 = (y_1 - y_3) + (y_2 - y_3).
\end{aligned}
\tag{4.38}
$$

Here, ξ_1 and ξ_2 describe the vertical displacements of the atoms Y_1 and Y_2 relative to the central atom X (atom 3) and play the role of the stretching modes in the linearised sense. The two coordinates ξ_3 and ξ_4 describe the bending mode and the orientation of the molecular plane about the z axis via the vector $(\Delta x, \Delta y) = (\xi_3, \xi_4)$ as shown in Fig. 4.7. These are two equivalent modes and are therefore degenerate in the description of the internal degrees of freedom of a linear molecule.

These coordinates ξ_i are in fact a linearisation of the following four geometrically defined coordinates:

$$
\begin{aligned}
\xi_1^{\rm GDC} =& r_1, \\
\xi_2^{\rm GDC} =& r_2, \\
\xi_3^{\rm GDC} =& -r_1 \left(\boldsymbol{e}_y \cdot \frac{\boldsymbol{r}_{13} \times \boldsymbol{r}_{23}}{r_{13}\, r_{23}} \right), \\
\xi_4^{\rm GDC} =& \ r_2 \left(\boldsymbol{e}_x \cdot \frac{\boldsymbol{r}_{13} \times \boldsymbol{r}_{23}}{r_{13}\, r_{23}} \right),
\end{aligned}
\tag{4.39}
$$

where coordinates $\xi_3^{\rm GDC}$ and $\xi_4^{\rm GDC}$ are defined using the normal to the molecule plane

$$
\boldsymbol{n} = \frac{\boldsymbol{r}_{13} \times \boldsymbol{r}_{23}}{r_{13}\, r_{23}},
$$

projected onto the coordinate axes x and y.

With the definition of the generalised vibrational degrees of freedom in Eq. (4.38), the linearised $\boldsymbol{B}$-matrix elements defining the gradients of $\boldsymbol{\xi}$ are then given by (only non-zero elements are shown)

$$
\begin{aligned}
B_{1,1,z} &= 1, & B_{1,3,z} &= 1, \\
B_{2,2,z} &= -1, & B_{2,3,z} &= 1, \\
B_{3,3,x} &= -2, & B_{4,3,2} &= -2, \\
B_{3,1,x} &= 1, & B_{3,2,y} &= 1, \\
B_{4,1,y} &= 1, & B_{4,2,y} &= 1.
\end{aligned}
$$

The corresponding $\boldsymbol{A}$-matrix elements are obtained by solving Eqs. (2.61–2.63) and are given by (for non-zero only)

$$\begin{aligned} 2A_{1,x,3} &= 2A_{1,y,4} = A_{1,z,1} = A_{1,z,2} = \\ 2A_{2,x,3} &= 2A_{2,y,4} = -A_{2,z,1} = -A_{2,z,2} = \\ = A_{3,x,3} &= A_{3,y,4} = A_{3,z,1} = A_{3,z,2} = \frac{\sqrt{m_Y}}{2\sqrt{m_Y}+\sqrt{m_X}}. \end{aligned}$$

Following the Sørensen $\boldsymbol{C}$-vector approach, the constraints defining the Eckart conditions and the equilibrium position have the following simple forms (non-zero elements only):

$$c_{x,2,y} = -c_{x,3,y} = -c_{y,2,x} = c_{y,3,x} = -m_Y\, r_{\rm e}.$$

The important consequence of the $3N-5$ approach is that the z components disappear from any elements representing the kinetic energy construction. For example, $\boldsymbol{J}$ is a 2×2 matrix:

$$\boldsymbol{J} = \begin{pmatrix} m_Y\, r_{12}\, r_{\rm e} & 0 \\ 0 & m_Y\, r_{12}\, r_{\rm e} \end{pmatrix},$$

where

$$r_{12} \equiv r_{\rm e} + \xi_1 + r_{\rm e} + \xi_2$$

is the vertical (along z) distance between atoms 1 and 2. The $\boldsymbol{J}$-matrix is diagonal, so is its inverse matrix $\boldsymbol{\eta} = \boldsymbol{J}^{-1}$:

$$\boldsymbol{\eta} = \begin{pmatrix} \frac{1}{r_{12}\, r_{\rm e}\, m} & 0 \\ 0 & \frac{1}{r_{12}\, r_{\rm e}\, m} \end{pmatrix}.$$

The construction of $\boldsymbol{s}^{\rm rot}$ and $\boldsymbol{s}^{\rm vib}$ via Eqs. (3.41) and (3.85) is straightforward and not reproduced here. Suffice to say is that none of the KEO elements are singular. For example, the determinant of the $\boldsymbol{t}$-matrix is given by

$$\det \boldsymbol{t} = -\frac{1}{4} r_{12}^2,$$

which is not zero unless $\xi_1 = r_{\rm e}$ and $\xi_2 = r_{\rm e}$, i.e. when three atoms coincide.

For the KEO elements, we obtain

$$G^{\rm rot}_{x,x} = G^{\rm rot}_{y,y} = \frac{1}{m_Y r_{12}^2},$$

$$\begin{aligned} G^{\rm vib}_{1,1} = G^{\rm vib}_{2,2} &= \frac{1}{\mu_{XY}} + \frac{1}{2m_Y r_{12}^2}\left[\xi_3^2+\xi_4^2\right], \\ G^{\rm vib}_{1,2} = G^{\rm vib}_{2,1} &= -\frac{1}{m_X} - \frac{1}{2m_Y r_{12}^2}\left[\xi_3^2+\xi_4^2\right], \\ G^{\rm vib}_{3,3} = G^{\rm vib}_{4,4} &= \frac{4}{\mu_{XY}} - \frac{8}{m_Y r_{12}^2}(r_{\rm e}+\xi_1)(r_{\rm e}+\xi_2), \end{aligned}$$

and

$$G_{1,x}^{\mathrm{Cor}} = -G_{2,x}^{\mathrm{Cor}} = -\frac{\xi_4}{m_Y\, r_{12}^2},$$
$$G_{1,y}^{\mathrm{Cor}} = -G_{2,y}^{\mathrm{Cor}} = \frac{\xi_3}{m_Y\, r_{12}^2},$$
$$G_{3,y}^{\mathrm{Cor}} = -G_{4,x}^{\mathrm{Cor}} = 2\frac{\xi_2 - \xi_1}{m_Y\, r_{12}^2},$$

where the following notation of the reduced mass is used:

$$\mu_{XY} = \frac{1}{m_X} + \frac{1}{m_Y}.$$

Finally, the pseudo-potential function vanishes,

$$U(\boldsymbol{\xi}) = 0.$$

As we have shown, the $3N-5$ formalism is fully consistent with the KEO construction formalism described above for the $3N-6$ case and can be performed using the same implementation of the $\boldsymbol{t}$–$\boldsymbol{s}$ formalism. The important difference is that the rotational degrees of freedom are contracted to x and y, with the z component (and $\hat{J}_z$) completely excluded from the calculation protocol.

Despite the very simple, non-singular form of the rectilinear KEO, such a non-singular ($3N-5$) rectilinear formalism is not very popular in modern variational calculations of high-resolution spectra. This is to a large extent because the rectilinear coordinates are less optimal to represent molecular potential energy functions. It is more efficient to design ro-vibrational basis sets that are capable of compensating singularities in the KEO. We will consider this approach as part of the basis sets chapter below.

MATERIALS USED

KEOs for triatomic molecules can be found in many works including Hougen (1962a), Howard and Moss (1971), Carter and Handy (1982), Carter et al. (1983), Sutcliffe and Tennyson (1986), Tennyson (1986), Sutcliffe and Tennyson (1991), Handy (1987) and Watson (1993), using in particular atom-diatom scattering coordinates, see Tennyson and Sutcliffe (1983) and Tennyson (1986). The Radau coordinates (Radau, 1868) were explained in Johnson and Reinhardt (1986) and used by Wei and Carrington (1997) and Wei and Jr. (1998) to derive an exact KEO for a triatomic molecule. An example of rectilinear coordinates for a linear molecule KEO was given in Wang et al. (2000).

The singularity of KEO at $\alpha = 0$ is discussed in Sutcliffe and Tennyson (1991). The triatomic $3N-6$ singularity treatment is from Yurchenko and Mellor (2020).

PAS constraints are from Sørensen (1979).

Hougen (1962a) and Howard and Moss (1971) showed how the normal commutation relations for $\hat{\boldsymbol{J}}$ can be restored via a suitable unitary transformation of the eigenfunctions, when the $3N-5$ approach is used for a linear molecule.

NOT COVERED BUT USEFUL MATERIALS

Hyperspherical coordinates KEO were developed by Johnson (1983) and Carter and Meyer (1990, 1994) and used for H_3^+.

CHAPTER 5

Basis sets

Basis sets are building blocks of the variational methodology. In this chapter, some key aspects of the construction and usage of rotational-vibrational basis sets are introduced. Here we follow the finite basis-set representation (FBR) approach, which assumes a set of predefined basis functions. The matrix elements in the FBR basis are computed via direct multi-dimensional integrals, which are most efficiently evaluated when both the basis functions and Hamiltonian operator are represented in a sum-of-products form. In this chapter different generalised approaches for the construction of product-form FBR basis sets are introduced with the emphasis on their numerical construction, accounting and usage. Different aspects of the matrix element calculations for sum-of-products Hamiltonian operators are also discussed. We show how to choose basis functions to resolve singularities of the KEO at linear geometries. The concept of assignment of ro-vibrational states based on their largest basis set contribution is introduced. Spherical harmonics are commonly used as rotational basis functions. Their basic properties and associated matrix elements of the angular momentum operators are described.

5.1 MATRIX ELEMENTS OF HAMILTONIAN

Our method of choice to solve the time-independent Schrödinger equation is the variational approach. It consists of representing eigenfunctions of the Hamiltonian operator

$$\hat{H} = \hat{T} + V$$

as linear combinations of appropriate basis functions with eigen coefficients obtained by diagonalising the Hamiltonian matrix $\hat{H}$ in that basis set. The kinetic energy operator (KEO) $\hat{T}$ is as in Eq. (3.5), which has a quadratic differential form, containing derivatives of up to the second order, and $V = V(\boldsymbol{\xi})$ is a potential energy function of the vibrational coordinates $\boldsymbol{\xi} = \{\xi_1, \xi_2, \ldots, \xi_M\}$.

It is possible and often useful to formulate the corresponding matrix elements of the second-order differential operators in terms of integrals of the first-order derivatives using integration by parts and taking advantage of the KEO form in Eq. (3.5) as follows. When evaluating integrals of the kinetic energy part $\hat{T}$ in

DOI: 10.1201/9780429154348-5

Eq. (3.5) involving $\hat{p}_\lambda = -i\hbar\partial/\partial\xi_\lambda$, with the help of integration by parts, the vibrational momentum operator $\hat{p}_\lambda$ is redirected to the function on the left side of the operator:

$$\begin{aligned}
&- i\hbar \int_a^b \phi(\xi_\lambda)\, \frac{\partial}{\partial \xi_\lambda}\, G_{\lambda,i}(\boldsymbol{\xi})\, \hat{\Pi}_i\, \phi'(\xi_\lambda)\, d\xi_\lambda = \\
&= -i\hbar\, \phi(\xi_\lambda)\, G_{\lambda,i}(\boldsymbol{\xi})\, \hat{\Pi}_i\, \phi'(\xi_\lambda)\Big|_a^b + i\hbar \int_a^b \frac{\partial \phi(\xi_\lambda)}{\partial \xi_\lambda}\, G_{\lambda,i}(\boldsymbol{\xi})\, \hat{\Pi}_i\, \phi'(\xi_\lambda)\, d\xi_\lambda = \\
&= i\hbar \int_a^b \frac{\partial \phi(\xi_\lambda)}{\partial \xi_\lambda}\, G_{\lambda,i}(\boldsymbol{\xi})\, \hat{\Pi}_i\, \phi'(\xi_\lambda)\, d\xi_\lambda = - \int_a^b \phi(\xi_\lambda)\, \hat{p}_\lambda^{\leftarrow}\, G_{\lambda,i}(\boldsymbol{\xi})\, \hat{\Pi}_i^{\rightarrow}\, \phi'(\xi_\lambda)\, d\xi_\lambda,
\end{aligned} \tag{5.1}$$

where $G_{\lambda,i}(\boldsymbol{\xi})$ are kinetic energy factors and the first term on the second line vanishes due to the bound state boundary conditions assumed here. We also introduce the following left-arrow notation for the vibrational momentum to indicate that it applies to the function on the left:

$$\phi(\xi_\lambda)\hat{p}_\lambda^{\leftarrow} = -i\hbar\frac{\partial \phi(\xi_\lambda)}{\partial \xi_\lambda}.$$

The right-arrow symbols in Eq. (5.1) are used to indicate the direction of the differential operator $\hat{\Pi}_i^{\rightarrow}$ acting on the function $\phi'(\xi_\lambda)$ to the right. Here $\hat{\Pi}_i$ is a generalised momentum operator, $\hat{\boldsymbol{\Pi}} = \{\hat{p}_1, \hat{p}_2, \hat{p}_3, \hat{J}_x, \hat{J}_y, \hat{J}_z\}$. In this form, only first-order derivatives of the wavefunctions are required. Analogously, for the vibrational diagonal KEO terms, we obtain

$$\begin{aligned}
&- \hbar^2 \int_a^b \phi(\xi_\lambda)\, \frac{\partial}{\partial \xi_\lambda}\, G_{\lambda,\lambda}(\boldsymbol{\xi})\, \frac{\partial}{\partial \xi_\lambda}\, \phi'(\xi_\lambda)\, d\xi_\lambda = \\
&= \hbar^2 \int_a^b \frac{\partial \phi(\xi_\lambda)}{\partial \xi_\lambda}\, G_{\lambda,\lambda}(\boldsymbol{\xi})\, \frac{\partial \phi'(\xi_\lambda)}{\partial \xi_\lambda}\, d\xi_\lambda = - \int_a^b \phi(\xi_\lambda)\, \hat{p}_\lambda^{\leftarrow}\, G_{i,\lambda}(\boldsymbol{\xi})\, \hat{p}_\lambda^{\rightarrow}\, \phi'(\xi_\lambda)\, d\xi_\lambda.
\end{aligned} \tag{5.2}$$

The matrix elements of the potential energy function V, the rotational KEO terms $\hat{J}_\alpha\, G^{\mathrm{rot}}_{\alpha,\beta}$ and pseudo-potential function U are of course derivative-free and do not require any special treatment.

5.2 RO-VIBRATIONAL BASIS SET

A ro-vibrational basis set is commonly built as a product of vibrational $\phi_v(\boldsymbol{\xi})$ and rotational $\phi^{(J)}_{k,m}$ basis functions:

$$\phi^{(J)}_{k,m,v} = \phi_v(\boldsymbol{\xi})|J,k,m\rangle, \tag{5.3}$$

where $|J,k,m\rangle$ is a rigid rotor wavefunction and v is a generic index referring to the vibrational part of the basis function. The rigid rotor (spherical harmonic) functions are by far the most optimal choice for a rotational basis set providing a complete set (see Section 5.6). Here k and m are the rotational quantum numbers associated with the projections of the angular momentum on the body-frame z and LF frame Z axes, respectively.

The ro-vibrational Hamiltonian in Eq. (3.5) can be factorised into a vibrational part $\hat{H}_{\text{vib}}$

$$\hat{H}_{\text{vib}} = \sum_{\lambda=1}^{M} \sum_{\lambda'=1}^{M} \hat{p}_{\lambda}\, G^{\text{vib}}_{\lambda,\lambda'}(\boldsymbol{\xi})\, \hat{p}_{\lambda'} + U(\boldsymbol{\xi}) + V(\boldsymbol{\xi}), \tag{5.4}$$

a Coriolis part

$$\hat{H}_{\text{Cor}} = \sum_{\alpha=x,y,z} \sum_{n=1}^{3N-6} \left[\hat{J}_{\alpha}\, G^{\text{Cor}}_{\alpha,\lambda}(\boldsymbol{\xi})\, \hat{p}_{\lambda} + \hat{p}_{\lambda}\, G^{\text{Cor}}_{\alpha,\lambda}(\boldsymbol{\xi})\, \hat{J}_{\alpha} \right]$$

and a rotational part

$$\hat{H}_{\text{rot}} = \frac{1}{2} \sum_{\alpha=x,y,z} \sum_{\alpha'=x,y,z} \hat{J}_{\alpha}\, G^{\text{rot}}_{\alpha,\alpha'}(\boldsymbol{\xi})\, \hat{J}_{\alpha'}.$$

The vibrational Hamiltonian $\hat{H}_{\text{vib}}$ is diagonal in the rotational basis set.

Matrix elements of the ro-vibrational terms of the Hamiltonian operator are then constructed by combining vibrational matrix elements of $G_{\alpha,\beta}(\boldsymbol{\xi})$ with matrix elements of the angular momentum operators $\hat{J}_{\alpha}$ and $\hat{J}_{\alpha}\hat{J}_{\beta}$ (described in detail in Section 5.6) as given by

$$\begin{aligned}\langle \phi^{(J)}_{k,m,v} | \hat{H} | \phi^{(J)}_{k',m,v'} \rangle &= \frac{1}{2} \sum_{\alpha,\alpha'=x,y,z} \langle J,k,m | \hat{J}_{\alpha}\hat{J}_{\beta} | J,k',m \rangle \, \langle v | G^{\text{rot}}_{\alpha,\beta} | v' \rangle \\ &- \frac{i\hbar}{2} \sum_{\alpha=x,y,z} \sum_{n=1}^{3N-6} \langle J,k,m | \hat{J}_{\alpha} | J,k',m \rangle \left[\langle v | G^{\text{Cor}}_{\alpha,n} \frac{\partial}{\partial \xi_n} + \frac{\partial}{\partial \xi_n} G^{\text{Cor}}_{\alpha,n} | v' \rangle \right] \\ &+ \delta_{k,k'} \langle v | H_{\text{vib}} | v' \rangle, \end{aligned} \tag{5.5}$$

where $\langle v| = \phi_v(\boldsymbol{\xi})^*$ and $|v'\rangle = \phi_{v'}(\boldsymbol{\xi})$. The rotational matrix elements do not depend on m because of space isotropy, and therefore the Hamiltonian matrix is diagonal both in J and m.

The Hamiltonian matrix is then diagonalised for an individual value of J to obtain the ro-vibrational energies (as eigenvalues) and the expansion eigen coefficients $C^{(J,\Gamma)}_{i,\lambda}$ (eigen coefficients), which in turn define the ro-vibrational eigenfunctions as a truncated basis set expansion

$$\Psi^{(J)}_{\lambda} = \sum_{k,v_1,\ldots,v_M} C^{(J)}_{i,\lambda} |J,k,m\rangle |v\rangle.$$

It is also desirable to take advantage of the symmetry of the system and use symmetrised basis functions that transform as one of the irreducible representations Γ of the symmetry groups that best describe the nuclear motion of the molecule. Symmetry-adapted representations of the Hamiltonian assume block-diagonal forms

$$\langle \Psi^{J,\Gamma}_{v,k} | \hat{H} | \phi^{J,\Gamma'}_{v,k} \rangle = H_{vk,v'k'} \delta_{J,J'} \delta_{\Gamma,\Gamma'}, \tag{5.6}$$

which means in practice that each (J, Γ)-block can be diagonalised independently with J and Γ as 'good quantum numbers' (i.e. constants of motion). The symmetrisation aspect of the basis sets is discussed in Chapter 6.

5.3 VIBRATIONAL BASIS SETS

5.3.1 Product-form basis sets

Traditionally vibrational basis sets are constructed as products of functions, e.g.

$$\phi_{\boldsymbol{k}}(\boldsymbol{\xi}) = \phi_{k_1}(\xi_1)\phi_{k_2}(\xi_2)\phi_{k_3}(\xi_3)\ldots \tag{5.7}$$

where $\boldsymbol{k} = \{k_1, k_2, k_3, \ldots\}$ are excitation indices and $\boldsymbol{\xi} = \{\xi_1, \xi_2, \xi_3, \ldots\}$ are some internal coordinates. This form provides an efficient integration of components of the Hamiltonian in Eq. (1.4), especially if they are also represented as sum-of-products, e.g.

$$\hat{H} = \sum_i h_{i_1}(\xi_1)h_{i_2}(\xi_2)h_{i_3}(\xi_3)\ldots$$

with the corresponding 1D integrals evaluated either analytically or numerically. Popular analytic basis sets include harmonic oscillators, Legendre polynomials and Laguerre polynomials, most of which are representative of the family of orthogonal polynomials $f_n(x)$ defined as

$$\int_a^b W(x)f_n(x)f_m(x)dx = 0 \quad (m \neq n),$$

where x is a coordinate ('abscissa') in the range $[a, b]$ and $W(x)$ is a positive weight function.

A very important property of orthogonal polynomials is that their integration can be efficiently formulated using the so-called Gaussian quadratures. Provided that an arbitrary function $g(x)$ is a finite linear combination of orthogonal polynomials $f_i(x)$ of the orders $i \leq M$, the following identity holds

$$\int_a^b W(x)g(x)dx = \sum_{i=1}^{M} w_i g(x_i),$$

where the abscissae x_i are (known) zeros of the orthogonal polynomial $f_M(x)$ and w_i are (known) weights. The power of this method for evaluating 1D integrals for the Hamiltonian matrix elements,

$$\int_a^b W(x)f_i(x)g(x)f_j(x)dx,$$

is that $f_i(x)g(x)f_j(x)$ do not have to be represented as combinations of orthogonal polynomials explicitly in order for the Gaussian quadratures to be applied:

$$\int_a^b W(x)f_i(x)g(x)f_j(x)dx \approx \sum_n w_n f_i(x_n)g(x_n)f_j(x_n), \tag{5.8}$$

where it is implicitly assumed that $g(x)$ can be represented as combinations of orthogonal polynomials. Thus the integrals are approximated by Gaussian quadratures, in which the accuracy can in principle be controlled by the number of quadrature points.

In fact, the discrete set of points (coordinates) and associated weights can be used as a basis set, thus forming the so-called discrete variable representation (DVR). In this book we concentrate on the FBR in the product form of Eq. (5.7). This form, before any improvements (such as symmetrisation or contraction, discussed below), will be also referred to as a 'primitive' basis set.

In principle a wealth of standard, well-tabulated orthogonal basis sets exist for vibrational basis set purposes, including orthogonal polynomials, Fourier series and many other finite basis functions. However, aiming at a purely numerical implementation, we do not have to be limited by analytically defined basis functions and may as well take advantage of the flexibility offered by numerical methods to generate orthogonal functions that are more optimal for the problem to be solved. Indeed, optimised wavefunctions can be produced by solving a simplified model eigenvalue problem for a reduced Hamiltonian as follows.

Let us consider a vibrational problem defined by $\hat{H}^{\rm vib}$ in Eq. (5.4) with M vibrational degrees of freedom. A 1D reduced Hamiltonian operator for a selected coordinate ξ_i can be constructed by setting all other $M-1$ coordinates to their equilibrium values to have the following general form:

$$H^{\rm 1D}(\xi_i) = -\frac{\hbar^2}{2}\frac{\partial}{\partial \xi_i} g_i \frac{\partial}{\partial \xi_i} + f(\xi_i) + u(\xi_i), \tag{5.9}$$

where g_i, f and u are $G_{i,i}$, V and U, respectively, with all other coordinates except ξ_i set to equilibrium. The corresponding one-dimensional vibrational Schrödinger equation

$$H^{\rm 1D}\phi_{n_i}(\xi_i) = E_{n_i}\phi_{n_i}(\xi_i) \tag{5.10}$$

can then be solved variationally. An efficient alternative to the variational methodology is the Numerov-Cooley approach, which is based on 'propagation of the solution from two boundaries'. In the case of equivalent vibrational modes, such as two stretching vibrations in XY_2, only one of each needs to be considered.

Figure 5.1 shows examples of 1D potential energy functions of PH_3, stretching $V(r)$ and bending $V(\alpha)$, constructed for one of the corresponding equivalent modes r_i or α_i ($i = 1, 2, 3$) from a 6D potential energy function $V(r_1, r_2, r_3, \alpha_1, \alpha_2, \alpha_3)$ by setting all other degrees of freedom to their equilibrium values. Here r_i are the valence bond lengths and α_i are the inter-bond angles as in Fig. 2.17. The stretching 1D potential function $V^{\rm 1D}(r)$ has a typical Morse-like shape and produces eigenfunctions with longer tails on the right-hand, dissociative side of the potential. As an illustration, one of the stretching eigenfunctions of $H^{\rm 1D}$ in Eq. (5.9), $\phi^{\rm str}_{v=6}(r)$ is shown. The bending potential function $V^{\rm 1D}(\alpha)$ has a typical less-unharmonic form with a more symmetric eigenfunction. As an example, $\phi^{\rm bnd}_{v=12}(\alpha)$ is shown ($\alpha \equiv \alpha_1$). The corresponding excitation number v is also the number of nodes of

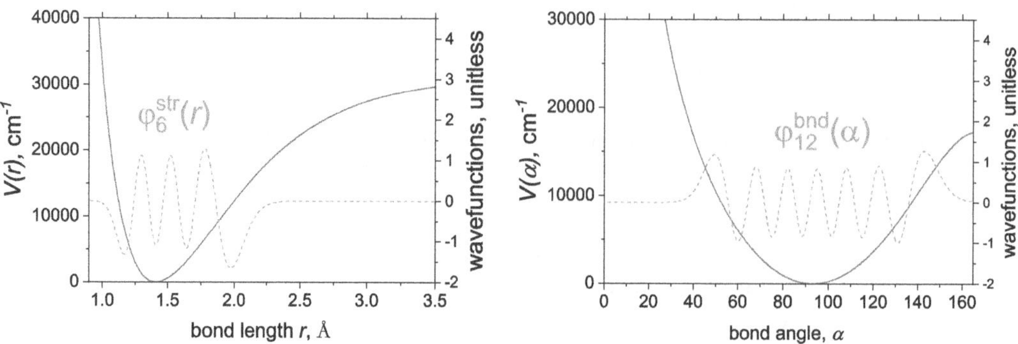

Figure 5.1 1D potential energy functions (solid lines) and selected wavefunctions (dashed lines) for PH_3. Functions $V(r)$ and $\phi_v^{\rm str}(r)$ (left) correspond to one of the three equivalent stretching modes r_1, r_2 and r_3. Functions $V(\alpha)$ and $\phi_v^{\rm bnd}(\alpha)$ correspond to one of the equivalent bending modes α_1, α_2 and α_3 (right). Selected wavefunctions are $\phi_6^{\rm str}(r)$ and $\phi_{12}^{\rm bnd}(\alpha)$ obtained as solutions of the corresponding Schrödinger equations in Eq. (5.10).

ϕ_v. The kinetic factor g_i is usually a slow function of g_i, at least for rigid molecules like PH_3.

Using these eigenfunctions, a full 6D primitive vibrational basis set is introduced as a product of 1D wavefunctions:

$$\phi_{v_1,v_2,v_3,v_4,v_5,v_6} = \phi_{v_1}^{\rm str}\phi_{v_2}^{\rm str}\phi_{v_3}^{\rm str}\phi_{v_4}^{\rm bnd}\phi_{v_5}^{\rm bnd}\phi_{v_6}^{\rm bnd}, \tag{5.11}$$

where $\phi_{v_i} = \phi_{v_i}(\xi_i)$ are eigenfunctions given in Eq. (5.9).

A practical example of a sum-of-product form of the Hamiltonian operator is the Taylor-type expansion as introduced for the KEO representation in Section 3.7. As a prerequisite of the full-D Hamiltonian matrix, only the following 1D primitive matrix elements for all ξ_i-dependent terms used in expansions of the Hamiltonian terms are required

$$\langle v_i|\xi_i^k|v_i'\rangle = \int_{a_i}^{b_i} \phi_{v_i}^* \xi_i^k \phi_{v_i'}\, d\xi_i, \tag{5.12}$$

$$\langle v_i|\frac{\partial}{\partial \xi_i}\xi_i^k\frac{\partial}{\partial \xi_i}|v_i'\rangle = -\int_{a_i}^{b_i} \frac{\partial \phi_{v_i}^*}{\partial \xi_i}\, \xi_i^k\, \frac{\partial \phi_{v_i'}}{\partial \xi_i}\, d\xi_i, \tag{5.13}$$

$$\langle v_i|\xi_i^k\frac{\partial}{\partial \xi_i}|v_i'\rangle = \int_{a_i}^{b_i} \phi_v^* \xi_i^k\, \frac{\partial \phi_{v_i'}}{\partial \xi_i}\, d\xi_i, \tag{5.14}$$

which can be easily evaluated numerically for 1D primitive basis functions defined either numerically or analytically. Here we use the bra-ket notation $|v_i\rangle = \phi_{v_i}(\xi_i)$ and $\langle v_i| = \phi_{v_i}^*(\xi_i)$. With the integrated-by-parts form of the KEO in Eq. (5.1), the following important identity for the integral in Eq. (5.14) holds

$$\langle v_i|\frac{\partial}{\partial \xi_i}\xi_i^k|v_i'\rangle = -\langle v_i'|\xi_i^k\frac{\partial}{\partial \xi_i}|v_i\rangle.$$

Moreover, there is no need for the second derivatives of the primitive basis function because of the property shown in Eq. (5.2).

The full-D primitive matrix elements of the Hamiltonian on the product basis in Eq. (5.7) are then constructed from 1D integrals as follows. The matrix elements of V are given by

$$\langle v_1, v_2, v_3, \ldots |V| v_1', v_2', v_3', \ldots\rangle = \sum_{i_1,i_2,i_3\ldots} f_{i_1,i_2,i_3\ldots} \prod_{k=1}^{M} \langle v_k|\xi_k^{i_k}|v_k'\rangle \tag{5.15}$$

assuming the Taylor-expanded form of the potential energy function

$$V(\xi_1, \xi_2, \xi_3, \ldots) = \sum_{i_1,i_2,i_3\ldots} f_{i_1,i_2,i_3\ldots}\xi_k^{i_k}.$$

The corresponding matrix elements of the Taylor-expanded kinetic energy terms U, $G_{\alpha,\beta}$, $G_{\lambda,\lambda'}$ and $G_{\lambda,\alpha}$ are given by

$$\langle \boldsymbol{v}|U|\boldsymbol{v}'\rangle = \sum_{i_1,i_2,i_3\ldots} u_{i_1,i_2,i_3\ldots} \prod_{k=1}^{M} \langle v_k|\xi_k^{i_k}|v_k'\rangle, \tag{5.16}$$

$$\langle \boldsymbol{v}|G_{\alpha,\beta}|\boldsymbol{v}'\rangle = \sum_{i_1,i_2,i_3\ldots} g^{\alpha,\beta}_{i_1,i_2,i_3\ldots} \prod_{k=1}^{M} \langle v_k|\xi_k^{i_k}|v_k'\rangle, \tag{5.17}$$

$$\langle \boldsymbol{v}|\frac{\partial}{\partial\xi_\lambda}G_{\lambda,\lambda}\frac{\partial}{\partial\xi_\lambda}|\boldsymbol{v}'\rangle = -\sum_{i_1,i_2,i_3\ldots} g^{\lambda,\lambda}_{i_1,i_2,i_3\ldots}\langle v_\lambda|\frac{\partial^{\leftarrow}}{\partial\xi_\lambda}\xi_\lambda^{i_\lambda}\frac{\partial^{\rightarrow}}{\partial\xi_\lambda}|v_\lambda'\rangle \prod_{k\neq\lambda} \langle v_k|\xi_k^{i_k}|v_k'\rangle, \tag{5.18}$$

$$\langle \boldsymbol{v}|\frac{\partial}{\partial\xi_\lambda}G_{\lambda,\lambda'}\frac{\partial}{\partial\xi_{\lambda'}}|\boldsymbol{v}'\rangle = \sum_{i_1,i_2,i_3\ldots} g^{\lambda,\lambda'}_{i_1,i_2,i_3\ldots}\langle v_\lambda|\hat{\pi}_\lambda|v_{\lambda'}\rangle\langle v_\lambda|\hat{\pi}_{\lambda'}|v_\lambda'\rangle \prod_{k\neq\lambda,\lambda'} \langle v_k|\xi_k^{i_k}|v_k'\rangle, \tag{5.19}$$

$$\langle \boldsymbol{v}|\frac{\partial}{\partial\xi_\lambda}G_{\lambda,\alpha} + G_{\alpha,\lambda}\frac{\partial}{\partial\xi_\lambda}|\boldsymbol{v}'\rangle = \sum_{i_1,i_2,i_3\ldots} g^{\lambda,\alpha}_{i_1,i_2,i_3\ldots}\langle v_\lambda|\hat{\pi}_\lambda|v_\lambda'\rangle \prod_{k\neq\lambda} \langle v_k|\xi_k^{i_k}|v_k'\rangle, \tag{5.20}$$

respectively. In these equations, the left/right derivatives are defined as in Eq. (5.1) and the operator $\hat{\pi}_\lambda$

$$\hat{\pi}_\lambda = \frac{\partial^{\leftarrow}}{\partial\xi_\lambda}\xi_\lambda^{k_\lambda} - \xi_\lambda^{k_\lambda}\frac{\partial^{\rightarrow}}{\partial\xi_\lambda}$$

was introduced for the sake of compactness.

For example, the (real and symmetric) matrix elements of the pure vibrational Hamiltonian $\hat{H}^{\rm vib}$ $(J = 0)$ are given by

$$\langle \boldsymbol{v}|H|\boldsymbol{v}'\rangle = -\frac{\hbar^2}{2}\sum_{\lambda,\lambda'}\langle \boldsymbol{v}|\frac{\partial}{\partial\xi_\lambda}G_{\lambda,\lambda}\frac{\partial}{\partial\xi_\lambda'}|\boldsymbol{v}'\rangle + \langle \boldsymbol{v}|U|\boldsymbol{v}'\rangle + \langle \boldsymbol{v}|V|\boldsymbol{v}'\rangle \tag{5.21}$$

and can be diagonalised on the basis of $\phi_{\boldsymbol{k}}(\boldsymbol{\xi})$ to obtain the $J = 0$ energies and vibrational wavefunctions.

5.3.2 Two-dimensional basis functions for doubly degenerate vibrations

For vibrational coordinates that are degenerate and therefore not (symmetry) independent, 1D wavefunctions are not always the best choice. For example, for the two deformation bending coordinates ξ_4 and ξ_5 of NH_3 in Eq. (2.80)

$$\begin{aligned} \xi_4 &= \frac{1}{\sqrt{6}}(2\alpha_1 - \alpha_2 - \alpha_3), \\ \xi_5 &= \frac{1}{\sqrt{2}}(\alpha_2 - \alpha_3), \end{aligned} \tag{5.22}$$

a preferable choice is to use 2D wavefunctions that correctly transform according to the degenerate irreducible representation E' of $\mathcal{D}_{3\mathrm{h}}(\mathrm{M})$, such as basis functions of a 2D isotropic harmonic oscillator (IHO).

The 2D IHO wavefunctions $|n_\mathrm{b}, l_\mathrm{b}, \tau_\mathrm{b}\rangle$ are obtained as solution of the Schrödinger equation

$$H^{\mathrm{IHO}}|n_\mathrm{b}, l_\mathrm{b}, \tau_\mathrm{b}\rangle = \omega_\mathrm{b}(n_\mathrm{b}+1)|n_\mathrm{b}, l_\mathrm{b}, \tau_\mathrm{b}\rangle,$$

for the IHO Hamiltonian given by

$$H^{\mathrm{IHO}} = \frac{1}{2}\left[\hat{P}_4^2 + \hat{P}_5^2 + \omega_\mathrm{b}^2\left(Q_4^2 + Q_5^2\right)\right],$$

where $\omega_\mathrm{b} = \sqrt{2 f^0_{4a4a} G^0_{4a,4a}}$ is the vibrational angular frequency of the 2D IHO and the IHO Hamiltonian operator is expressed in terms of normal-mode coordinates $Q_4 = \xi_4, \quad Q_5 = \xi_4$ and their conjugate momenta.

The eigenvalues of H^{IHO} are independent of l_b so that an energy level with a given value of n_b is $(n_\mathrm{b}+1)$-fold degenerate. Here the 2D IHO is defined in terms of polar coordinates $\{r_\mathrm{b}, \varphi_\mathrm{b}\}$ related to ξ_4 and ξ_5 as follows:

$$\begin{aligned} r_\mathrm{b}\cos(\varphi_\mathrm{b}) &= \left(\frac{2f^\mathrm{e}_{4,4}}{G^\mathrm{e}_{4,4}}\right)^{1/4} \times \xi_4, \\ r_\mathrm{b}\sin(\varphi_\mathrm{b}) &= \left(\frac{2f^\mathrm{e}_{4,4}}{G^\mathrm{e}_{4,4}}\right)^{1/4} \times \xi_5, \end{aligned}$$

where $G^\mathrm{e}_{4,4}$ is the equilibrium value of kinetic energy factor $G_{4,4}$, f^e_{44} is the second derivative of the potential energy function V with respect to ξ_4 (at equilibrium) and we assume that for the IHO, $G^\mathrm{e}_{4,4} = G^\mathrm{e}_{5,5}$ and $f^\mathrm{e}_{4,4} = f^\mathrm{e}_{5,5}$.

The IHO wavefunctions are then given by

$$\begin{aligned} |n_\mathrm{b}, l_\mathrm{b}, 0\rangle &= N_{n_\mathrm{b},l_\mathrm{b}} \exp(-r_\mathrm{b}^2/2)\, r_\mathrm{b}^{l_\mathrm{b}}\, L^{l_\mathrm{b}}_{(n_\mathrm{b}-l_\mathrm{b})/2}(r_\mathrm{b}^2)\, \cos(l_\mathrm{b}\,\varphi_\mathrm{b}), \\ |n_\mathrm{b}, l_\mathrm{b}, 1\rangle &= N_{n_\mathrm{b},l_\mathrm{b}} \exp(-r_\mathrm{b}^2/2)\, r_\mathrm{b}^{l_\mathrm{b}}\, L^{l_\mathrm{b}}_{(n_\mathrm{b}-l_\mathrm{b})/2}(r_\mathrm{b}^2)\, \sin(l_\mathrm{b}\,\varphi_\mathrm{b}), \end{aligned} \tag{5.23}$$

where $L_k^{l_\mathrm{b}}$ is an associated Laguerre polynomial, n_b is the principal vibrational quantum number and l_b is the vibrational angular momentum quantum number; $|n_\mathrm{b}, l_\mathrm{b}, 0\rangle$ has positive parity, while $|n_\mathrm{b}, l_\mathrm{b}, 1\rangle$ is a negative-parity function. Here we use the real IHO eigenfunctions in terms of $\cos(l_\mathrm{b}\,\varphi_\mathrm{b})$ and $\sin(l_\mathrm{b}\,\varphi_\mathrm{b})$, rather than the usual complex form involving $\exp(il_\mathrm{b}\,\varphi_\mathrm{b})$ in order to keep the Hamiltonian matrix real. The vibrational angular momentum l_b can only take positive values $n_\mathrm{b}, n_\mathrm{b}-2, \ldots, 1$ or 0. The normalisation constant is

$$N_{n_\mathrm{b},l_\mathrm{b}} = \sqrt{2-\delta_{l_\mathrm{b},0}}\sqrt{\frac{[(n_\mathrm{b}-l_\mathrm{b})/2]!}{\pi\,[(n_\mathrm{b}+l_\mathrm{b})/2]!}}. \tag{5.24}$$

The full-D vibrational basis set for NH_3 ($J=0$) is then given by

$$\phi_\nu(\boldsymbol{\xi}) = \phi_{n_1}(\xi_1)\,\phi_{n_2}(\xi_2)\,\phi_{n_3}(\xi_3)\,\phi_{n_\mathrm{b},l_\mathrm{b},\tau_\mathrm{b}}(r_\mathrm{b},\varphi_\mathrm{b})\,\phi_{n_6}(\xi_6), \tag{5.25}$$

where $\phi_{n_k}(\xi_k) \equiv |n_k\rangle$ ($k = 1,2,3,6$) are some 1D primitive basis functions and coordinates ξ_i are as in Section 2.6.2.

While isotropic 2D wavefunctions are desirable, in principle it is also possible to use a product of two 1D oscillators for the description of the degenerate modes, which might be advantageous for practical numerical implementations. The drawback of such a representation is the complex permutation properties of corresponding 1D components. As an illustration, let us consider a bending basis function of NH_3 given by

$$\phi_{v_4,v_5}(\xi_4,\xi_5) = \phi_{v_4}(\xi_4)\phi_{v_4}(\xi_5)$$

and apply a permutation (123) of the three equivalent nuclei H_i. The three bending coordinates α_1, α_2 and α_3 transform upon the permutation operation (123) as follows:

$$(123)\begin{pmatrix}\alpha_1\\ \alpha_2\\ \alpha_3\end{pmatrix} = \begin{pmatrix}\alpha_3\\ \alpha_1\\ \alpha_2\end{pmatrix}.$$

This transformation mixes the corresponding degenerate pair (ξ_4,ξ_5) in Eq. (5.22) as

$$(123)\begin{pmatrix}\xi_4\\ \xi_5\end{pmatrix} = \begin{pmatrix}-\frac{1}{2}\xi_4 - \frac{\sqrt{3}}{2}\xi_5\\ \frac{\sqrt{3}}{2}\xi_4 - \frac{1}{2}\xi_5\end{pmatrix}.$$

As a result, the transformation breaks the product form of primitive bending functions, i.e.

$$(123)\phi_{n_4}(\xi_4)\phi_{n_5}(\xi_5) = \phi_{n_4}\left(-\frac{1}{2}\xi_4 + \frac{\sqrt{3}}{2}\xi_5\right)\phi_{n_5}\left(-\frac{\sqrt{3}}{2}\xi_4 - \frac{1}{2}\xi_5\right),$$

which cannot be directly expressed in a compact form of products of $\phi_{n_4}(\xi_4)\phi_{n_5}(\xi_5)$ and $\phi_{n_4}(\xi_5)\phi_{n_5}(\xi_4)$ unless involving combinations of, at least in principle, infinite number of other basis set components. An exception is the product basis $\phi_{n_4}\phi_{n_5}$ of

1D harmonic oscillator wavefunctions, which can be shown to transform correctly using a finite set of functions belonging to the same principal quantum number $n_{\rm b} = n_4+n_5$. Indeed, a product of two degenerate 1D harmonic oscillator functions $\phi_{v_4,v_5}(\xi_4,\xi_5)$

$$\phi^{\rm HO}_{n_4,n_5}(\xi_4,\xi_5) = C_{n_4,n_5} H_{n_4}(\xi_4) e^{-\alpha\xi_4^2} H_{n_5}(\xi_5) e^{-\alpha\xi_5^2}$$

is also a solution of a 2D degenerate harmonic oscillator hamiltonian

$$\hat{H}^{\rm IHO}\phi^{\rm HO}_{n_4,n_5}(\xi_4,\xi_5) = \omega_{\rm b}(n_4+n_5+1)\phi^{\rm HO}_{n_4,n_5}(\xi_4,\xi_5).$$

Therefore, $\phi^{\rm HO}_{n_4,n_5}$ can be directly related to 2D IHO wavefunctions as linear combinations of $\phi_{n_{\rm b},l_{\rm b},\tau_{\rm b}}(r_{\rm b},\varphi_{\rm b})$ with the same $n_{\rm b} = n_4+n_5$ and different $l_{\rm b}$ and thus also have the compact transformational properties of IHO. Here H_n is a Hermite polynomial of order n, α is a parameter (the same for all degenerate components) and C_{n_4,n_5} is a normalisation constant.

For example, for $n_{\rm b} = n_4+n_5 = 3$, the following four products $\phi_{n_4}(q_4)\phi_{n_5}(q_5)$

$$\phi_3\phi_0, \quad \phi_2\phi_1, \quad \phi_1\phi_2, \quad \phi_0\phi_3 \tag{5.26}$$

form a full set of harmonic functions with simple and separable permutation properties:

$$(123)\begin{pmatrix}\phi_3\phi_0\\ \phi_2\phi_1\\ \phi_1\phi_2\\ \phi_0\phi_3\end{pmatrix} = \boldsymbol{D}[(123)]\begin{pmatrix}\phi_3\phi_0\\ \phi_2\phi_1\\ \phi_1\phi_2\\ \phi_0\phi_3\end{pmatrix},$$

where $\boldsymbol{D}[(123)]$ is a 4×4 matrix. Like IHO, these four wavefunctions are degenerate and share the same harmonic oscillator energy

$$\tilde{E}^{\rm HO}_{n_4,n_5} = \tilde{\omega}\,(n_4+n_5+1).$$

A more detailed discussion of symmetry properties of vibrational basis functions is given in Chapter 6.

5.3.3 3D vibrational basis functions

Using the same logic, a 3D isotropic vibrational basis set can be constructed and used for triply degenerate vibrational modes. For example, the F-symmetry bending or stretching coordinates of CH_4 represent triply degenerate vibrational modes, such as in Eqs. (2.100) or (2.101), respectively. The appropriate isotropic, triply degenerate basis functions in this case are the spherical harmonics, or Wigner D functions, defined as follows. The corresponding degenerate coordinates ξ_x, ξ_y and ξ_z are expressed in the spherical polar form as

$$\begin{pmatrix}\xi_x\\ \xi_y\\ \xi_z\end{pmatrix} = r_s\begin{pmatrix}\sin\theta_s\cos\phi_s\\ \sin\theta_s\sin\phi_s\\ \cos\theta_s\end{pmatrix}, \tag{5.27}$$

where $r_s^2 = \xi_x^2 + \xi_y^2 + \xi_z^2$. The 3D isotropic basis functions are then given by

$$\phi_{n,l,m} = N_{n,l}\, r_s^l\, e^{-\nu r_s^2}\, L_n^{l+1/2}(\sqrt{\nu} r_s)\, Y_{l,m}(\theta_s, \phi_s),$$

where N_{kl} is a normalisation constant. Here $Y_{l,m}(\theta_s, \phi_s)$ is a spherical harmonic function, l is the vibrational angular momentum quantum number, m is its projection along the molecule-fixed z axis and $L_n^{l+1/2}(2\nu r_s^2)$ is an associated Laguerre polynomials with k as the radial excitation number and $\nu = \frac{\mu\omega}{2\hbar}$, where μ is the reduced mass of the system.

The wavefunction $\phi_{k,l,m}$ is a solution of the Schrödinger equation for the 3D potential

$$V(r_s) = \frac{1}{2}\mu\omega^2 r_s^2$$

with the energy given by

$$E_n = \hbar\omega\left(n + \frac{3}{2}\right), \qquad n = 0, 1, 2, 3, \ldots$$

Analogously to the 2D case, it is also possible to use a triple product of 1D harmonic oscillator functions to represent the triply degenerate basis set with the advantage of having a more universal and therefore robust implementation of the basis set: the same construction algorithm based on 1D basis set functions can be used for any vibrational modes regardless of their nature, degeneracy or symmetry properties. The advantage of the native 2D or 3D representation is in their more natural transformation properties.

5.3.4 Basis set pruning: polyad and energy thresholds

The curse of the variational approach is the exponential growth in the size of the basis set with the dimensionality of the vibrational degrees of freedom. Indeed, the size of the raw product-form vibrational basis set is $N_1 \times N_2 \times, \ldots, \times N_M$ (a volume of a hypercube) with N_i as the number of basis functions for mode i. The rotational basis set for each value of J contributes with a factor of $2J+1$ due to the rotational basis functions $|J, k\rangle$ with k ranging from $-J$ to J. Even for tetratomic polyatomic molecules, the product form of the basis set quickly becomes intractable. For example, taking a relatively small 6D vibrational basis set with $v_i = 0, 1, 2, 3, 4$ (five functions each), the total size of the direct product is $5^6 = 15625$, which for $J = 30$ leads to a matrix of 953 125×953 125 elements representing a challenge for a direct diagonalisaton.

It is therefore important to use some way of pruning the vibrational basis set without significant loss of accuracy. One popular choice is to use polyad and energy thresholds. Consider a product-form vibrational basis set for the PH_3 molecule, constructed from harmonic oscillator wavefunctions, stretching and bending. For PH_3, we define the polyad number as given by

$$P = 2(n_1 + n_3) + n_2 + n_4, \tag{5.28}$$

where n_1, n_2, n_3 and n_4 are the normal-mode quantum numbers used to describe the six vibrational degrees of freedom of PH_3 (with ν_3 and ν_4 as 2D degenerate modes). This polyad formulation is constructed to roughly describe how vibrational energies depend on quantum numbers assuming the harmonic approximation. For the case of the vibrations of PH_3, the harmonic energies are characterised by four harmonic constants $\tilde{\omega}_1$ (symmetric stretch), $\tilde{\omega}_2$ (symmetric bend), $\tilde{\omega}_3$ (asymmetric, double degenerate stretch) and $\tilde{\omega}_4$ (asymmetric, double degenerate bend) which approximately relate as

$$\tilde{\omega}_1 \approx \tilde{\omega}_3 \approx 2\tilde{\omega}_2 \approx 2\tilde{\omega}_4.$$

In this approximation, the vibrational states of n_1, n_2, n_3 and n_4 corresponding to the same polyad number P are degenerate with the energy

$$E_{n_1,n_2,n_3,n_4} = \tilde{\omega}_1\left(n_1 + \frac{1}{2}\right) + \tilde{\omega}_2\left(n_2 + \frac{1}{2}\right) + \tilde{\omega}_3(n_3+1) + \tilde{\omega}_4(n_4+1) = \tilde{\omega}\left(P + \frac{9}{2}\right),$$

thus forming energy clusters for different P, defined in Eq. (5.28). In practice, the vibrational energies of PH_3 are quite different from those of the harmonic oscillator, but this approximation is still good enough to reflect the polyad structure of the vibrational energies, as shown in Fig. 5.2. A polyad clustering structure suggests an efficient mechanism to control the size of the basis set by selecting an appropriate polyad threshold $P_{\max}$, $P \le P_{\max}$, which provides a simple, one-parameter pruning of the vibrational basis set. For example, by setting a threshold condition to

$$P = 2(n_1 + n_3) + n_2 + n_4 \le P_{\max} = 8,$$

where $n_1, n_3 \le 4$ and $n_2, n_4 \le 8$. The bending modes sample twice as high excitations as the stretching modes. With the condition $2(n_1 + n_3) + n_2 + n_4 \le 8$, the

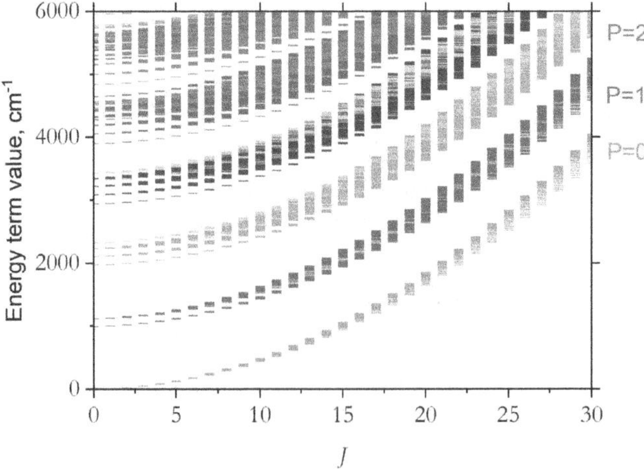

Figure 5.2 PH_3 energy polyad structure.

size of the vibrational basis is reduced to only 742, compared to the total number of combinations of the direct product basis of $5^3 \times 9^3 = 91125$.

5.3.5 Basis set bookkeeping

In numerical implementations involving (Hamiltonian) matrices representing multi-dimensional basis functions and vectors, it is important to associate multi-dimensional index space $\{v_1, v_2, \ldots, v_M\}$ with a single index v, i.e. $v \leftrightarrow \{v_1, v_2, \ldots, v_M\}$.

There are many ways by which the mapping between v and $\{v_1, v_2, \ldots, v_M\}$ for the product form of the of the primitive basis set

$$|v\rangle = |v_1, v_2, \ldots, v_M\rangle = |v_1\rangle|v_2\rangle \ldots |v_M\rangle$$

may be organised. Here we present a mapping scheme based on the multi-dimensional summation approach in Eq. (3.57) introduced in Section 3.7.1 to formalise the construction of Taylor-type coordinate expansions. With the polyad number P playing the role of the expansion order L in Eq. (3.57), we aim to arrange states $|v_1, v_2, \ldots, v_M\rangle$ in increasing P order and assign a counting index v to the corresponding combinations $\{v_1, v_2, \ldots, v_M\}$.

Let us consider as an example a vibrational basis set for PH_3 in the product form of 1D basis functions as in Eq. (5.11). For this curvilinear coordinate representation, the polyad number P in terms of the corresponding vibrational excitation numbers v_i is given by

$$P = 2(v_1 + v_2 + v_3) + v_4 + v_5 + v_6. \tag{5.29}$$

Using the polyad structured summation, the linear expansion of variationally obtained eigenfunctions in terms of the primitive basis functions

$$\Phi_i = \sum_{v_1,v_2,v_3,v_4,v_5,v_6} C_{v_1,v_2,v_3,v_4,v_5,v_6}|v_1\rangle|v_2\rangle|v_3\rangle|v_4\rangle|v_5\rangle|v_6\rangle = \sum_{\boldsymbol{v}} C_{\boldsymbol{v}}|\boldsymbol{v}\rangle \tag{5.30}$$

is arranged into the following sum:

$$\Phi_i = \sum_{P=0}^{P_{\max}} \sum_{v_1=0}^{P/2} \sum_{v_2=0}^{P/2-v_1} \sum_{v_3=0}^{P/2-v_1-v_2} \sum_{v_4=1}^{P-2(v_1-v_2-v_3)} \sum_{v_5=0}^{P-2(v_1-v_2-v_3)-v_4} \times C_{v_1,v_2,v_3,v_4,v_5,v_6}|v_1\rangle|v_2\rangle|v_3\rangle|v_4\rangle|v_5\rangle|v_6\rangle,$$

with the constraint for v_6 of

$$v_6 = P - 2(v_1 + v_2 + v_3) - v_4 - v_5.$$

The size of the basis set is defined as the summation limit $v_{\max}$ and is obtained as the total number of elements in the sum

$$v_{\max} = \sum_{P=0}^{P_{\max}} \sum_{i_1=0}^{P/2} \sum_{i_2=0}^{P/2-i_1} \sum_{i_3=0}^{P/2-i_1-i_2} \sum_{i_4=0}^{P-2(i_1-i_2-i_3)} \sum_{i_5=0}^{P-2(i_1-i_2-i_3)-i_4} 1.$$

Table 5.1 A polyad mapping of a 6D primitive index $\{v_1, v_2, \ldots, v_6\}$ and a 1D index v for the polyad number defined as $P = 2(v_1 + v_2 + v_3) + v_4 + v_5 + v_6$

v	P	v_1	v_2	v_3	v_4	v_5	v_6
1	0	0	0	0	0	0	0
2	1	0	0	0	0	0	1
3	1	0	0	0	0	1	0
4	1	0	0	0	1	0	0
5	2	0	0	0	0	0	2
6	2	0	0	0	0	1	1
7	2	0	0	0	0	2	0
8	2	0	0	0	1	0	1
9	2	0	0	0	1	1	0
10	2	0	0	0	2	0	0
11	2	0	0	1	0	0	0
12	2	0	1	0	0	0	0
13	2	1	0	0	0	0	0
...							

An example of basis set mapping $|v\rangle \leftrightarrow |v_1\rangle|v_2\rangle|v_3\rangle|v_4\rangle|v_5\rangle|v_6\rangle$ for $P \le 2$ is illustrated in Table 5.1. The total number of primitive basis set functions in this case is $v_{\rm max} = 13$.

In a numerical implementation, the mapping between the index v and quantum numbers $\{v_1, v_2, v_3, v_4, v_5, v_6\}$ can be obtained by counting from $v = 1$ for $\{0, 0, 0, 0, 0, 0\}$ using the following pseudo-code:

```
v ← 0
for P = 0, P_max do
    for i1 = 0, P_max/2 do
        for i2 = 0, P_max/2 − i1 do
            for i3 = 0, P_max/2 − i1 − i3 do
                for i4 = 0, P_max − 2(i1 − i2 − i3) do
                    for i5 = 0, P_max − 2(i1 − i2 − i3) − i4 do
                        i6 ← P_max − 2(i1 − i2 − i3) − i4 − i5
                        v ← v + 1
                        if {i1, i2, i3, i4, i5, i6} ≡ {v1, v2, v3, v4, v5, v6} then
                            return i
                        end if
                    end for
                end for
            end for
        end for
    end for
end for
```

The rigid six-loop structure in this example of the basis set mapping is for the specific case of six degrees of freedom specifically. A more flexible implementation of the mapping, applicable for arbitrary systems of $M = 3N - 6$ degrees of freedom, can be constructed by adopting the recursive procedure INDEX from Section 3.7.1, introduced there for Taylor-like expansions of a Hamiltonian.

As an alternative to the polyad scheme, an energy threshold $\tilde{E}_{\max}$ can be used to control the basis set truncation. In this case, an approximate energy value $\tilde{E}_{v_1,\ldots,v_6}$ is associated with the basis function $|v_1\rangle,\ldots,|v_6\rangle$ as, e.g., a sum of 1D contributions $\tilde{E}_{v_i}$:

$$\tilde{E}^{\text{approx.}}_{v_1,\ldots,v_6} = \tilde{E}_{v_1} + \tilde{E}_{v_2} + \tilde{E}_{v_3} + \tilde{E}_{v_4} + \tilde{E}_{v_5} + \tilde{E}_{v_6},$$

where $\tilde{E}_{v_1}$ is an eigenvalue of the corresponding 1D Hamiltonian. The (product-form) basis set is then constructed to include all combinations $|v_1\rangle|v_2\rangle\ldots|v_M\rangle$ satisfying the condition

$$\tilde{E}^{\text{approx.}}_{v_1,\ldots,v_M} \le \tilde{E}_{\max}.$$

For example, by applying $\tilde{E}_{\max} = 8000$ cm^{-1} to the harmonic basis set of PH_3 described above and assuming $\tilde{\omega}_1 = \tilde{\omega}_3 = 2424.4$ cm^{-1} and $\tilde{\omega}_2 = \tilde{\omega}_4 = 1100.5$ cm^{-1}, the primitive basis set is reduced to 316 wavefunctions. An energy truncation is more flexible than the polyad scheme, but at the same time less robust. The number of basis functions (i.e. the size of the basis set) satisfying the energy threshold criteria exhibits a non-trivial dependence on the value $\tilde{E}_{\max}$ and is therefore more difficult to predict.

As should be clear by now, a reduction of the basis set plays a crucial role in ro-vibrational calculations of polyatomic molecules. In the following, we consider two approaches commonly used to optimise the size of a multi-dimensional basis set, the basis set contraction and symmetrisation.

5.4 BASIS SET CONTRACTION

5.4.1 The $J = 0$ representation

Since some basis sets are better suited for variational calculations than others, it is generally beneficial to find or even construct basis functions optimised for the specific physical problem considered. We expect that by choosing a more suitable basis set, the number of the basis functions can be reduced. In Section 5.3.1 we showed how to build tailored primitive 1D basis functions by solving specially constructed 1D model Schrödinger equations. In fact one can extend this idea to the next level of complexity and optimise primitive basis sets for higher dimensions by solving multi-dimensional model problems, tailored to the problem in question. One of such models, leading to a powerful vibrational basis set, is the so-called $J = 0$ contraction.

We start by solving the pure vibrational ($J = 0$) Schrödinger equation

$$\hat{H}^{J=0}\Phi_\lambda^{(J=0)} = E_\lambda^{(J=0)}\Phi_\lambda^{(J=0)} \tag{5.31}$$

in the product form of the primitive (orthogonal) basis functions $\phi_{v_1,v_2,\ldots,v_M}$ as in Eq. (5.7), where λ represents the state counting number. By construction, vibrational eigenfunctions $\Phi_\lambda^{(J=0)}$ are orthogonal to each other and thus can be utilised as basis functions for solving the ro-vibrational problem. Moreover, $\Phi_{\lambda}^{(J=0)}$ are expected to provide a better basis with a more compact structure than the original primitive basis functions $\phi_{v_1,v_2...}$ they are based on. Therefore, the size of the basis set can be also significantly reduced by omitting functions corresponding to higher excitations, thus justifying the '$J = 0$ contraction' name.

Ro-vibrational basis functions are then formed as products of $\Phi_{\lambda}^{(J=0)}$ and rigid rotor basis functions $|J, k, m\rangle$

$$\Phi_{\lambda,k,m,}^{(J)} = \Phi_{\lambda}^{(J=0)}|J, k, m\rangle. \tag{5.32}$$

The $J = 0$ contracted functions offer another useful advantage: the vibrational part of the Hamiltonian operator $\hat{H}^{\mathrm{vib}} \equiv \hat{H}^{(J=0)}$ in Eq. (5.5) is diagonal in $|\lambda\rangle = \Phi_{\lambda}^{(J=0)}$, which simplifies the calculation of matrix elements by replacing them with $E_{\lambda}^{(J=0)}\delta_{\lambda,\lambda'}$. Another, less obvious feature is that the calculated values $E_{\lambda}^{(J=0)}$ in this substitution can be replaced by more accurate, experimental values if available. This leads to effective shifts of the corresponding vibrational bands helping to improve rotational energies within a given vibrational state λ.

Consider our example of the $P = 8$ basis set of PH_3 with the 742 wavefunctions and energies ranging up to ~10000 cm^{-1}. By solving the vibrational Schrödinger equation given in Eq. (5.31), where we find 742 eigenfunctions $\Phi_{\lambda}^{(J=0)}$. We now apply an energy cut-off of 8000 cm^{-1} to contract to 443 wavefunctions, representing a smaller and more compact basis set than original $\phi_{v_1,...,v_6}$ towards the ro-vibrational basis set $\Phi_{\lambda,k,m,}^{(J)}$ in Eq. (5.32). The dimension of the contracted ro-vibrational basis set for $J = 30$ is 27023, which is significantly smaller than the uncontracted ro-vibrational primitive basis of $91125 \times (2J + 1) = 2\ 733\ 750$ functions.

At this point, let us make a side note about the symmetry of the eigen-solution, which will be important later. The Hamiltonian operator $\hat{H}^{\mathrm{vib}}$ is totally symmetric and also commutes with all the operations of the symmetry group it belongs to (for example $C_{3\mathrm{v}}$ in the case of PH_3). Therefore, its eigenfunctions $\Psi_{\lambda}^{(0)}$ are also eigenfunctions of all symmetry operations of the group, i.e. they transform according to one of the irreducible representations (we just do not immediately know which one). The basis functions obtained by the $J = 0$ contraction scheme automatically form a symmetry-adapted basis set, which is a useful property provided effectively for free. We will explore this property below where symmetrisation of basis sets is discussed.

5.4.2 'Contracted' representation of Hamiltonian

Let us now consider $\Phi_{\lambda}^{(J=0)}$ as vibrational basis functions and formulate the Hamiltonian ro-vibrational matrix in this representation. Assuming the product form of the primitive functions $|i\rangle = \phi_{v_1,v_2,...,v_M}$ in Eq. (5.7), the $J = 0$ matrix representation of $\hat{H}$ can be obtained via a unitary transformation of the primitive matrix elements of the kinetic and energy terms; see Eqs. (5.15–5.20).

We first restructure the ro-vibrational Hamiltonian as follows:

$$\hat{H} = \frac{1}{2} \sum_{\alpha,\alpha'=x,y,z} \hat{J}_\alpha\, G^{\rm rot}_{\alpha,\alpha'}(\boldsymbol{\xi})\, \hat{J}_{\alpha'} - \frac{i\hbar}{2} \sum_{\alpha=x,y,z} \bar{G}^{\rm Cor}_\alpha(\boldsymbol{\xi}) \hat{J}_\alpha + \hat{H}^{\rm vib},$$

where

$$\hat{H}^{\rm vib} = -\frac{\hbar^2}{2} \sum_{\lambda,\lambda=1}^{M} \frac{\partial}{\partial \xi_\lambda} G^{\rm vib}_{\lambda,\lambda'}(\boldsymbol{\xi}) \frac{\partial}{\partial \xi'_\lambda} + U(\boldsymbol{\xi}) + V(\boldsymbol{\xi}),$$

$$\bar{G}^{\rm Cor}_\alpha(\boldsymbol{\xi}) = G^{\rm Cor}_{\alpha,\lambda}(\boldsymbol{\xi}) \frac{\partial}{\partial \xi_\lambda} + \frac{\partial}{\partial \xi_\lambda} G^{\rm Cor}_{\alpha,\lambda}(\boldsymbol{\xi}).$$

Remember that $\hat{H}^{\rm vib}$ is conveniently diagonal in the $J = 0$ representation (see Eq. (5.31)),

$$\langle \Phi^{(J=0)}_\lambda | \hat{H}^{\rm vib} | \Phi^{(J=0)}_{\lambda'} \rangle = E^{(J=0)}_\lambda \delta_{\lambda,\lambda'}$$

and therefore requires no additional work. The matrix elements of the kinetic energy terms $G^{\rm rot}_{\alpha,\alpha'}(\boldsymbol{\xi})$ and $\bar{G}^{\rm Cor}_\alpha(\boldsymbol{\xi})$ in this representation are obtained via a finite unitary transformation $\boldsymbol{T} = |\Phi^{(J=0)}\rangle\langle \mathbf{i}|$ appeared in the solution

$$|\Phi^{(J=0)}_\lambda\rangle = \sum_{i=1}^{i_{\rm max}} T_{i,\lambda} |i\rangle,$$

where i is the vibrational counting index as in Eq. (5.30). The matrix $\boldsymbol{T}$ consists of eigen coefficients of $\hat{H}^{(J=0)}$ in the primitive basis $|i\rangle$, satisfying the unitary condition

$$\boldsymbol{T}^\dagger \boldsymbol{T} = \mathbb{1},$$

where $\mathbb{1}$ is a unitary matrix.

The $J = 0$ representation of the rotational G-terms is then given by

$$\langle \Phi^{(J=0)}_\lambda | G^{\rm rot}_{\alpha,\beta} | \Phi^{(J=0)}_{\lambda'} \rangle = \sum_{i,i'} T^*_{i,\lambda} \langle i | G^{\rm rot}_{\alpha,\beta} | i' \rangle T_{i',\lambda'} \tag{5.33}$$

or

$$\boldsymbol{G}^{(J=0)}_{\alpha,\beta} = \boldsymbol{T}^\dagger \boldsymbol{G}^{\rm prim}_{\alpha,\beta} \boldsymbol{T}. \tag{5.34}$$

Similarly, for the Coriolis matrix elements, we obtain

$$\bar{\boldsymbol{G}}^{(J=0)}_\alpha = \boldsymbol{T}^\dagger \bar{\boldsymbol{G}}^{\rm prim}_\alpha \boldsymbol{T}. \tag{5.35}$$

Here $\boldsymbol{G}^{(J=0)}_{\alpha\ldots}$ and $\boldsymbol{G}^{\rm prim}_{\alpha\ldots}$ denote the $J = 0$ and primitive representations of the corresponding terms $G^{\rm rot}_{\alpha\beta}$ and $\bar{G}^{\rm Cor}_\alpha$, respectively.

The transformation matrices $\boldsymbol{T}$ are usually large. The evaluation of the matrix elements in Eqs. (5.34) and (5.35) is effectively a triple-matrix product of the type

$$A^{(J=0)}_{\lambda,\lambda'} = \sum_{i=1}^{i_{\rm max}} \sum_{i'=1}^{i_{\rm max}} T^*_{i,\lambda} B^{\rm prim}_{i,i'} T_{i',\lambda'} \tag{5.36}$$

requiring $i^3_{\rm max}$ operations and is therefore time and memory consuming.

A well-known trick to reduce the number of operations in the product of three matrices is to replace the two simultaneous summations by two consecutive matrix products:

$$C^{(\mathrm{inter})}_{i,\lambda'} = \sum_{i'=1}^{i_{\max}} B^{\mathrm{prim}}_{i,i'} T_{i',\lambda'} \tag{5.37}$$

and

$$A^{(J=0)}_{\lambda,\lambda'} = \sum_{i=1}^{i_{\max}} T^{*}_{i,\lambda} C^{(\mathrm{inter})}_{i,\lambda'}, \tag{5.38}$$

where $\boldsymbol{C}^{(\mathrm{inter})}$ is an intermediate matrix. It is easy to show that the two-step single summations in Eqs. (5.37, 5.38) 'cost' significantly less, $2i^{\,2}_{\max}$ operations, which is especially important for large $i_{\max}$.

5.4.3 Another layer of contraction

So far we were able to significantly reduce the size of the ro-vibrational basis set by dividing it into two subspaces based on their physical properties (vibrational and rotational) and contracting the vibrational part. We will now show how to exploit the 'divide-and-contract' methodology further and partition the vibrational basis set into specially selected subgroups of vibrational modes. There are different ways of selecting vibrational subspaces. For example, in (ro-)vibrational calculations of dimers (e.g. $(H_2O)_2$), it is natural to use the monomers' wavefunctions when forming dimer's vibrational basis set, where the closely coupled modes of monomers are combined onto independent subspaces. Here we explore the molecular symmetry argument and show how vibrational subspaces can be formed from equivalent degrees of freedom. This approach will also be useful later for the construction of symmetry-adapted basis functions.

Let us consider the PH_3 example again, with six valence vibrational coordinates defined in Section 2.6.1. The corresponding product-form primitive basis functions $\phi_{v_1,v_2,v_3,v_4,v_5,v_6}$ are given in Eq. (5.11) with three equivalent stretching $\phi^{\mathrm{str}}_{v_i}$ ($i = 1, 2, 3$) and three equivalent bending $\phi^{\mathrm{bnd}}_{v_i}$ ($i = 4, 5, 6$) basis functions. The corresponding degrees of freedom $\{r_1, r_2, r_3\}$ and $\{\alpha_1, \alpha_2, \alpha_3\}$ are fully independent in their symmetric properties and naturally form two subspaces: stretching subspace 1 and bending subspace 2. We now solve two reduced Schrödinger equations for subspaces 1 and 2

$$\begin{aligned} \hat{H}^{(1)}_{\mathrm{str}} \Psi^{(1)}_{\lambda_1} &= E_{\lambda_1} \Psi^{(1)}_{\lambda_1}, \\ \hat{H}^{(2)}_{\mathrm{bnd}} \Psi^{(2)}_{\lambda_2} &= E_{\lambda_2} \Psi^{(2)}_{\lambda_2}. \end{aligned} \tag{5.39}$$

Here we choose to construct the reduced 3D Hamiltonian operators $\hat{H}^{(1)}_{\mathrm{str}}$ and $\hat{H}^{(2)}_{\mathrm{bnd}}$ by vibrationally averaging the total vibrational Hamiltonian $\hat{H}^{\mathrm{6D}}$ over the ground state basis functions from subspace 2 and 1, respectively:

$$\begin{aligned}\hat{H}^{(1)}_{\rm str}(r_1, r_2, r_3) &= \langle 0_4|\langle 0_5|\langle 0_6|\hat{H}^{\rm 6D}|0_6\rangle|0_5\rangle|0_4\rangle,\\ \hat{H}^{(2)}_{\rm bnd}(\alpha_1, \alpha_2, \alpha_3) &= \langle 0_1|\langle 0_2|\langle 0_3|\hat{H}^{\rm 6D}|0_3\rangle|0_2\rangle|0_1\rangle,\end{aligned} \tag{5.40}$$

where r_i and α_i $(i = 1, 2, 3)$ are stretching and bending coordinates, respectively; $|0_n\rangle$ denotes ground state $(v_n = 0)$ basis functions $\phi^{\rm str}_{v_n}$ $(n = 1, 2, 3)$ or $\phi^{\rm bnd}_{v_n}$ $(n = 4, 5, 6)$.

Once eigenfunctions for each subspace $s = 1, 2$ are found, the final vibrational basis set is formed as a direct product

$$\Psi_{\lambda_1,\lambda_2} = \Psi^{(1)}_{\lambda_1} \otimes \Psi^{(2)}_{\lambda_2}.$$

Before solving the $J = 0$ Schrödinger equation (5.31), this basis set can be contracted by removing some functions $\Psi_{\lambda_1,\lambda_2}$ that are expected to be unimportant for the subsequent solution, e.g. either components corresponding to high excitation (polyad contraction) or components corresponding to very high energies. Analogously, more contraction layers can be introduced to simplify solutions of large systems.

Vibrational matrix elements of the Hamiltonian matrix $\hat{H}^{\rm 6D}$ on the contracted basis of $\Psi_{\lambda_1,\lambda_2}$ are now constructed through a product of unitary transformations

$$\boldsymbol{T}^{\rm contr} = \boldsymbol{T}^{(1)}\boldsymbol{T}^{(2)};$$

see Eqs. (5.34, 5.35). The transformation matrices $\boldsymbol{T}^{(1)}$ and $\boldsymbol{T}^{(2)}$ are eigen coefficients of $\Psi^{(1)}_{\lambda_1}$ and $\Psi^{(2)}_{\lambda_2}$, respectively:

$$\begin{aligned}\Psi^{(1)}_{\lambda_1} &= \sum_n T_{i,\lambda_1}|i_{\rm str}\rangle = \sum_{v_1,v_2,v_3} T^{\lambda_1}_{v_1,v_2,v_3}|v_1, v_2, v_3\rangle,\\ \Psi^{(2)}_{\lambda_2} &= \sum_n T_{i,\lambda_2}|i_{\rm bnd}\rangle = \sum_{v_4,v_5,v_6} T^{\lambda_2}_{v_4,v_5,v_6}|v_4, v_5, v_6\rangle\end{aligned}$$

in the primitive basis set representation.

Let us consider now a non-rigid molecule NH_3 with the vibrational coordinates ξ_i defined in Section 2.6.2 (three stretching, two symmetrised dihedral and one inversion mode) and form a 6D vibrational basis set using the reduced Hamiltonian scheme introduced above. Based on their properties, the six vibrational modes of NH_3 form three subspaces, stretching $\{\xi_1, \xi_2, \xi_3\}$, bending $\{\xi_4, \xi_5\}$, and inversion $\{\xi_6\}$, which are all symmetry independent from each other.

The product-type vibrational primitive basis set for NH_3 is then given by (see Section 5.3.2)

$$\phi_\nu(\boldsymbol{\xi}) = \phi_{n_1}(\xi_1)\phi_{n_2}(\xi_2)\phi_{n_3}(\xi_3)\phi_{n_4,n_5}(\xi_4, \xi_5)\phi_{n_6}(\xi_6), \tag{5.41}$$

where $\phi_{n_k}(\xi_k) \equiv |n_k\rangle$ ($k = 1, 2, 3$), $\phi_{n_4,n_5}(\xi_4, \xi_5) \equiv |n_4, n_5\rangle$ and $\phi_{n_6}(\xi_6) \equiv |n_6\rangle$ are some generic primitive basis functions. The three reduced Hamiltonian operators for each $i = 1, 2, 3$ coordinate subspace are then given by

$$\begin{aligned}
\hat{H}^{(1)}(\xi_1, \xi_2, \xi_3) &= \langle 0_4, 0_5|\langle 0_6|\hat{H}|0_6\rangle|0_4, 0_5\rangle, \\
\hat{H}^{(2)}(\xi_4, \xi_5) &= \langle 0_1|\langle 0_2|\langle 0_3|\langle 0_6|\hat{H}|0_6\rangle|0_3\rangle|0_2\rangle|0_1\rangle, \\
\hat{H}^{(3)}(\xi_6) &= \langle 0_1|\langle 0_2|\langle 0_3|\langle 0_4, 0_5|\hat{H}|0_4, 0_5\rangle|0_3\rangle|0_2\rangle|0_1\rangle,
\end{aligned}$$

where $|0\rangle_k$ are ground state vibrational basis functions $\phi_0(\xi_k)$ ($k = 1, \ldots, 6$). The corresponding three eigenvalue problems are given by

$$\hat{H}^{(i)}(\boldsymbol{\xi}^{(i)})\Psi^{(i)}_{\lambda_i}(\boldsymbol{\xi}^{(i)}) = E_{\lambda_s}\Psi^{(i)}_{\lambda_s}(\boldsymbol{\xi}^{(i)}). \tag{5.42}$$

As discussed above, we expect all eigenvectors of Eq. (5.42) to transform according to some irreducible representations $\Gamma^{(i)}_s$. Vibrational degrees of freedom of NH_3 can be described by irreducible representations of the molecular symmetry group $\mathcal{D}_{3h}$(M), which will be discussed in Chapter 6.

As a final example, consider a 9D problem of the XY_4-type tetrahedral molecule with the coordinates defined in Section 2.9. Based on their symmetry properties, the 9D coordinate space can be divided into three subspaces: (i) $\{\xi_1, \xi_2, \xi_3, \xi_4\}$ (stretching), (ii) $\{\xi_5, \xi_6\}$ (E-type bending) and (iii) $\{\xi_7, \xi_8, \xi_9\}$ (F-type bending). Each of the subspaces is symmetrically independent and thus can be processed separately. This partitioning can be used to solve the Schrödinger equations for each subspace using the corresponding basis functions: $\Phi^{(\mathrm{i})}_{\lambda_1}(\xi_1, \xi_2, \xi_3, \xi_4)$, $\Phi^{(\mathrm{ii})}_{\lambda_2}(\xi_5, \xi_6)$ and $\Phi^{(\mathrm{iii})}_{\lambda_3}(\xi_7, \xi_8, \xi_9)$. The final vibrational basis set is formed from a truncated direct product $\Psi^{(\mathrm{i})}_{\lambda_1} \otimes \Psi^{(\mathrm{ii})}_{\lambda_2} \otimes \Psi^{(\mathrm{iii})}_{\lambda_3}$.

5.4.4 Basis set contraction with the vibrational angular momentum

We showed how to build vibrational basis sets as eigenfunctions of $\hat{H}_i$ by solving an eigenvalue problem for a reduced Hamiltonian operator as in Eq. (5.39). In fact, $\hat{H}_i$ is not the only choice and any other suitable vibrational operators could be used. An example of such an alternative is the vibrational angular momentum operator $\hat{L}^2$. The vibrational basis set of $\hat{L}^2$ are classified with the vibrational angular momentum quantum number l, which is a useful property for some basis sets, including the IHOs. For the NH_3 example above, the reduced bending Hamiltonian problem (subspace 2) given by Eq. (5.42) can be replaced with the following equation of a vibrational angular momentum:

$$\hat{L}^2_{\mathrm{bnd}}\Psi^{(2)}_{\lambda_2,l_2} = \hbar^2 l_2(l_2+1)\Psi^{(2)}_{\lambda_2},$$

where $\hat{L}^2_{\mathrm{bnd}}$ is a 2D operator obtained by vibrationally averaging the total vibrational angular momentum $\hat{L}^2$ (see, e.g., Eq. (3.87)) over ground state basis functions $\phi_0(\xi_k)$ from subspaces 1 and 3:

$$\hat{L}^2_{\mathrm{bnd}}(\xi_4, \xi_5) = \langle 0_1|\langle 0_2|\langle 0_3|\langle 0_6|\hat{L}^2|0_6\rangle|0_3\rangle|0_2\rangle|0_1\rangle,$$

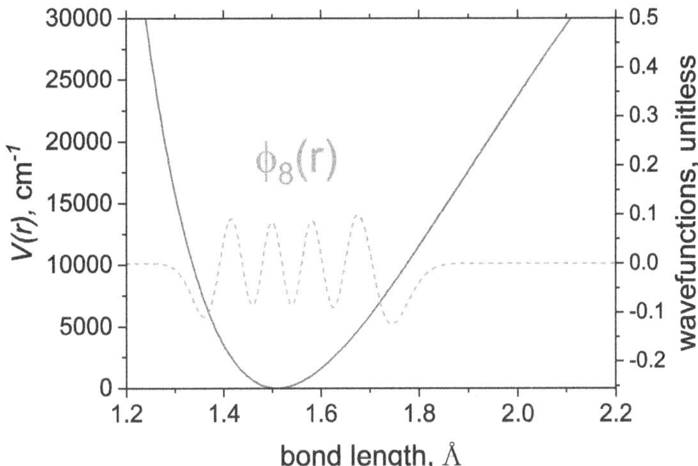

Figure 5.3 Potential energy function $V(r)$ (solid line) and an example of a vibrational eigenfunction $\phi_8(r)$ (dashed line) of SiO in its ground electronic state $X\ ^1\Sigma^+$.

where $|0\rangle_k = \phi_0(\xi_k)$ ($k = 1, 2, 3$ and 6). The corresponding eigenfunctions $\Psi^{(2)}_{\lambda_2,l_2}$ are obtained with l_2 as a good quantum number.

5.4.5 Assignment: vibrational quantum numbers

One important aspect of spectroscopic calculations is to provide quantum number assignments for computed states. Only two quantum quantities are generally rigourous in variational ro-vibrational calculations, the rotational angular momentum J and the total symmetry Γ (see Chapter 6); other quantum numbers are usually approximate. Let us first discuss how to assign variationally computed energies and wavefunctions with (approximate) vibrational quantum numbers v_i. Rotational quantum numbers are discussed in Section 5.6.

The assignment of 1D vibrational states (e.g. in the case of diatomics) can be generally established by counting nodes of the corresponding 1D (bound) eigenfunctions $\phi_v(\xi)$ (see, e.g., Fig. 5.1), at least for simple potential functions with a single minimum. Indeed, the vibrational energies in this case increase with the excitation number v, which is also the number of the nodes of $\phi_v(\xi)$. Figure 5.3 shows an example of a vibrational wavefunction of SiO in its ground electronic state for $v = 8$ with eight nodes.

In principle the assignment of multi-dimensional vibrational functions can be also associated with their crossings of the zero plane; however, these are significantly more difficult to count. Figure 5.4 illustrates three 2D stretching wavefunctions of H_2S obtained by solving a reduced 2D vibrational problem variationally using a

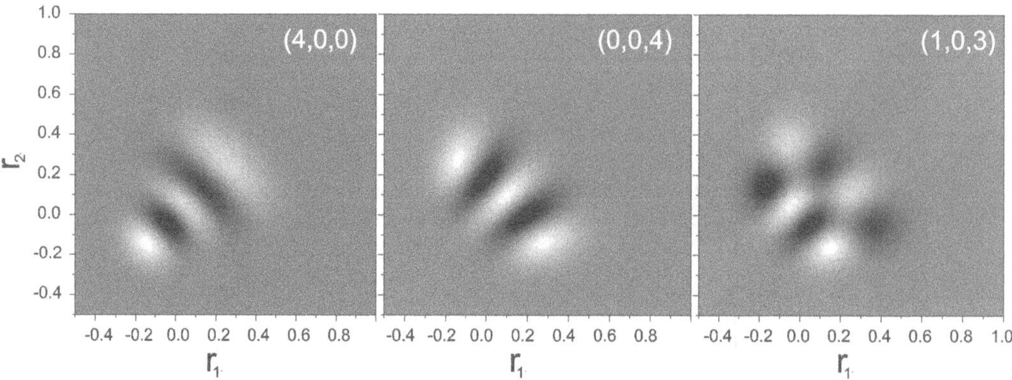

Figure 5.4 2D stretching vibrational wavefunctions $\phi_{v_1,v_2,v_3}(r_1, r_2)$ of H_2S representing three excited states (4,0,0), (0,0,4) and (1,0,3), v_1 is the symmetric stretch, v_2 is the symmetric bend and v_3 is asymmetric stretch quantum numbers.

product-type basis set

$$\phi_{v_1,v_2} = \phi_{v_1}(r_1)\phi_{v_2}(r_2)$$

with some appropriate stretching basis functions. For this illustration, a small basis set of 15 basis functions was constructed from excitations $v_1 = 0, 1, 2, 3, 4$ and $v_2 = 0, 1, 2, 3, 4$, capped with the polyad-type threshold condition

$$v_1 + v_2 \leq 4.$$

For 2D wavefunctions shown for this simplified case in Fig. 5.4, one can recognise their nodes and even count excitations: $\Psi_{4,0,0}$ has four nodes along the diagonal between r_1 and r_2 and is fully symmetric (A_1 in $\mathcal{C}_{2v}$); $\Psi_{0,0,4}$ is also fully symmetric with four nodes along the line perpendicular to that of $\Psi_{4,0,0}$; $\Psi_{1,0,3}$ is a mixture with one symmetric and three asymmetric excitations. Here we use the standard spectroscopic (normal mode) notations for the three vibrational modes, ν_1 is the symmetric stretch, ν_2 is the (symmetric) bend and ν_3 is the asymmetric stretch; see Fig. 2.15. For example, $\Psi_{0,0,4}$ is a four quanta excitation along the asymmetric mode ν_3.

Using visual inspections in state assignments is, however, not an option for real numerical applications characterised by high excitations, large number of states and higher dimensions. Therefore, it is more common to use approximate vibrational quantum numbers for polyatomic vibrational wavefunctions. Even approximate assignment is extremely useful for analysis of variational solutions and even more so for relating calculations to experiment or other calculations. The main concept behind assigning vibrational quantum numbers is to compare vibrational eigenfunctions ψ_i (or corresponding densities $|\psi_i|^2$) to and associate with some reference wavefunctions with well-defined quantum numbers, e.g. basis functions. Consider, for example, variationally computed wavefunctions of the form

$$\psi_i = \sum_{v_1,v_2,\ldots} C^i_{v_1,v_2,\ldots,}\phi_{v_1,v_2,v_3\ldots v_M}. \tag{5.43}$$

An eigen-state ψ_i is said to inherit quantum numbers $v'_1, v'_2, \ldots, v'_M$ of the most 'similar' basis function $\phi_{v'_1,v'_2,v'_3\ldots v'_M}$. The concept of similarity may be explored in different ways. For example, according to the popular largest contribution concept, the eigen-state ψ_i is assigned the excitation numbers $v'_1, v'_2, \ldots, v'_M$ that correspond to the largest basis set contribution measured by the corresponding expansion (eigen-)coefficient $|C^i_{v'_1,v'_2,\ldots,}|^2$. The accuracy of this approach depends on the quality of the basis set and can only provide meaningful information when the coefficient is relatively large (larger than 0.5 in accordance with the theorem of Hose and Taylor) and the coupling between different basis set components is relatively small, i.e. when the magnitude of the largest coefficients $|C^i_{v'_1,v'_2,\ldots,}|^2$ is well separated from other contributions. The magnitude of the largest coefficient plays a role of a similarity index between the eigenfunction and the corresponding reference function. It is also common for spectroscopic applications to relate the vibrational assignment to normal-mode quantum numbers with the associated harmonic normal-mode wavefunctions.

This approach can be illustrated using one of the wavefunctions of H_2S from Fig. 5.4. Solving the 2D Schrödinger equation variationally, i.e. in the form of Eq. (5.43), $\Psi_{1,0,3}$ was obtained as given by

$$\begin{aligned}\Psi^{B_2}_{1,0,3} =& 0.0042(|10\rangle - |01\rangle) + 0.0088(|20\rangle - |02\rangle) - 0.0055(|30\rangle - |03\rangle) + \\ & 0.14(|12\rangle - |21\rangle) + 0.41(|40\rangle - |40\rangle) - 0.55(|31\rangle - |13\rangle) + \ldots,\end{aligned} \tag{5.44}$$

where $|v_1 v_2\rangle = \phi_{v_1}(r_1)\phi_{v_2}(r_2)$.

It is clear that the largest basis set contributions are from the $v_1 + v_2 = 4$ components, which, together with the symmetry of the sate B_2, suggest either $(1, 0, 3)$ or $(3, 0, 1)$ as assignment (using the spectroscopic normal-mode convention). This is the maximum one can learn about the state quantum numbers without bringing additional arguments or analysis. For more on the assignment of the state symmetries, see Chapter 6.

An alternative to the largest contribution approach is to use an overlap with an appropriate set of orthonormal reference wavefunctions (e.g. harmonic oscillator) as the similarity index $|C^i_{v_1,v_2,\ldots,}|$

$$C^i_{v_1,v_2,\ldots,} = \langle v_1|\langle v_2|\ldots\langle v_M|\psi_i\rangle.$$

5.5 RESOLUTION OF SINGULARITIES AT LINEAR GEOMETRIES: NON-RECTILINEAR COORDINATES

5.5.1 Formulation

Previously (Section 4.4) we showed how to build non-singular KEOs for linear or quasi-linear triatomic molecules using a $3N - 5$ rectilinear coordinate frame.

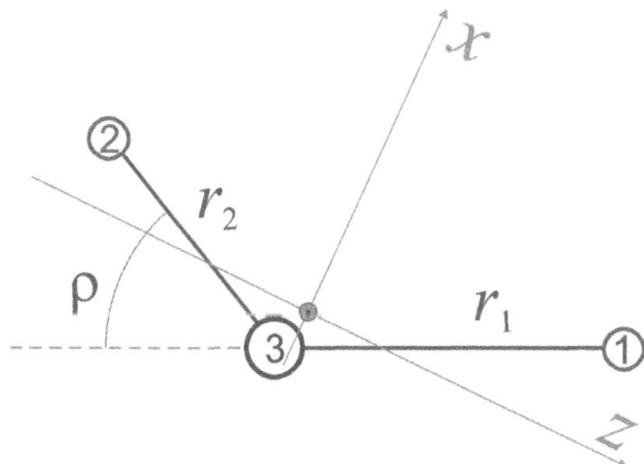

Figure 5.5 Body-fixed coordinates for an XY_2-type molecule with $\rho = 0$ at linear geometry.

In this Section we consider an alternative approach to resolving singularities of (quasi-)linear triatomic KEOs, based on specially constructed basis functions and the $3N-6$ coordinate representation. Instead of seeking a singularity-free coordinate transformation, we now accept the presence of a singularity in the $(3N-6)$ KEO and learn how to deal with it by using appropriate basis set functions. What matters for the variational methodology is that the KEO matrix elements must be finite. This is analogous to the hydrogen atom problem with the singular potential of $1/r$ and the corresponding radial wavefunctions vanishing at $r \to 0$ in the appropriate form so that all integrals are finite.

Let us return to the XY_2 example from Section 4.1 with the valence coordinates and bisector frame. The two bond lengths r_1 and r_2 represent two stretching vibrations, while for the bending mode we adopt the coordinate $\rho = \pi - \alpha$, with $\rho = 0$ at the linear geometry of $\alpha = \pi$ (Fig. 5.5) as in Section 3.7.3. The KEO is the same as derived in Section 4.1, see Eqs. (4.16), (4.20), (4.21) and (4.23), where we need to replace $\alpha \to \pi - \rho$ and $\partial/\partial\alpha \to -\partial/\partial\rho$. The KEO terms $G^{\rm rot}_{x,x}$, $G^{\rm rot}_{z,z}$, $G^{\rm rot}_{x,z}$ and U become singular at $\rho \to 0$. The pseudo-potential function U and the rotational KEO factor $G^{\rm rot}_{z,z}$ have a singularity of the type $\frac{1}{\sin^2\rho}$ ($\sim 1/\rho^2$), while the singularity of $G^{\rm rot}_{x,z}(r_1, r_2, \rho)$ is of the type $\frac{1}{\sin\rho}$ ($\sim 1/\rho$). The term $G^{\rm rot}_{x,x}(r_1, r_2, \rho)$ also becomes singular but at a different geometry, when the molecule is fully folded with $\rho = \pi$, which is of the type $\frac{1}{\cos^2(\rho/2)}$.

Effectively, a (quasi-)linear molecule will be treated as a bent molecule with the linear configuration accessible through molecular vibration. This is different from the $3N-5$ approach, which is effectively built around its linear configuration.

Among the most popular choices of the bending basis sets for resolving a singularity are the associated Legendre polynomials $P_n^k(x)$. The corresponding orthogonal basis functions constructed from the real normalised associated Legendre polynomials are given by

$$\psi_n^{(k)}(\rho) = \sqrt{\sin\rho}\,\bar{P}_n^{(k)}(\cos\rho), \tag{5.45}$$

with the normalisation condition

$$\int_{-1}^{1}\left[\bar{P}_n^{(k)}(x)\right]^2 dx = \int_{o}^{\pi}\left[\bar{P}_n^{(k)}(\cos\rho)\right]^2 \sin\rho\, d\rho = \int_{0}^{\pi}\left|\psi_n^{(k)}\right|^2 d\rho = 1,$$

where $x = \cos\rho$ and

$$\bar{P}_n^{(k)} \equiv \sqrt{\frac{(2l+1)!(l-k)!}{2(l+k)!}}P_n^{(k)}.$$

The rotational-angular basis ($J > 0$) is given by

$$\psi_{n,J,k,m}^{\rm rv} = \psi_n^{(k)}|J,k,m\rangle,$$

where k is both the rotational quantum number (projection of the angular momentum on the z axis) and the index of the associated Legendre polynomial.

Let us show that with the wavefunctions in Eq. (5.45) and kinetic energy $\boldsymbol{G}$ factors in Eqs. (4.11–4.23), all singularities cancel out. We first consider the pseudo-potential term, which we process together with the diagonal bending kinetic term $G_{33} \equiv G_{\rho,\rho}$:

$$\frac{\hbar^2}{2}\frac{\partial\psi_n^{(k)}}{\partial\rho}G_{3,3}\frac{\partial\psi_m^{(k)}}{\partial\rho} + \psi_n^{(k)}U\psi_m^{(k)} = \frac{\hbar^2}{2}\sin\rho\frac{\partial\bar{P}_n^{(k)}}{\partial\rho}G_{3,3}\frac{\partial\bar{P}_m^{(k)}}{\partial\rho} + \bar{P}_n^{(k)}\tilde{U}\bar{P}_m^{(k)}, \quad (5.46)$$

to obtain a new pseudo-potential term for this basis $\bar{P}_n^{(k)}$

$$\tilde{U} = \frac{\hbar^2}{8}\left[-2\cos\rho\frac{\partial G_{3,3}}{\partial\rho} + \left(2\sin\rho + \frac{\cos^2\rho}{\sin\rho}\right)G_{3,3}\right] + U\sin\rho = \hbar^2\frac{\sin(2\rho)}{2r_1r_2m_X}, \quad (5.47)$$

which has a very compact form and is not singular.

The singularity of the term $G_{z,z}\hat{J}_z^2$ in Eq. (4.23) is not relevant for $k = 0$ since it vanishes upon $\hat{J}_z$ acting on $|J,k,m\rangle$. For $k > 0$ the associated Legendre polynomials $\bar{P}_n^{(k)}$ can be always represented in the following form:

$$\bar{P}_n^{(k)}(\cos\rho) = \sin^k(\rho)\,\tilde{P}_n^{(k)}(\cos\rho). \quad (5.48)$$

The factor $\sin^k(\rho)$ then resolves the singularity in Eq. (4.23) for any $k \geq 1$ as follows:

$$\frac{\hbar^2}{2}G_{z,z}k^2\psi_n^{(k)}\psi_{n'}^{(k)} = \frac{\hbar^2}{2}\tilde{G}_{z,z}k^2\sin^{2k-1}(\rho)\,\tilde{P}_n^{(k)}(\cos\rho)\,\tilde{P}_{n'}^{(k)}(\cos\rho), \quad (5.49)$$

where $\psi_n^{(k)}$ is from Eq. (5.45) and

$$\tilde{G}_{z,z} = G_{z,z}\sin^2\rho = \cos^2(\rho/2)\left[\frac{1}{\mu_{XY}}\left(\frac{1}{r_1^2}+\frac{1}{r_2^2}\right) + \frac{2}{m_Xr_1r_2}\right], \quad (5.50)$$

which is also non-singular.

Finally, the $1/\sin\rho$ singularity of the $G_{z,x}$ term is analogously resolved through the leading term $\sin\rho$ in the definition of the basis function in Eq. (5.45).

The singularity of the $G_{x,x}$ term in Eq. (4.20) is a special case. It diverges at $\rho = \pi$ and cannot be resolved using the associated Legendre polynomials when $k = 0$ and $J > 0$. In practice, however, this geometry of a fully bent molecule ($\alpha = 0$) corresponds to a very high energy and thus is not very important. For example, before any integration, the primitive basis functions $\sqrt{\sin\rho}\bar{P}_n^{(k)}(\cos\rho)$ can be optimised by eigen-solving the pure angular reduced Hamiltonian for a realistic angular potential energy function $\bar{V}(\rho)$ as was discussed in Section 5.3.1. The optimised solution usually vanishes at the limit $\rho \to \pi$ faster than $1/\cos^2(\rho/2)$ in Eq. (4.20).

5.5.2 Associated Laguerre polynomials

As an alternative to the Legendre polynomial, the basis set can be formed from the associated Laguerre polynomials $L_n^{(l)}(\rho)$. Not only do these orthogonal polynomials allow for the full resolution of the $\rho = 0$ singularity in Eqs. (4.16, 4.21 and 4.23), they also provide a more compact, physically motivated basis set. Laguerre bending basis functions, constructed from the (real) associated Laguerre polynomials $L_n^{(l)}(\rho)$, are given by

$$\psi_n^{(l)}(\rho) = C_{n,l}\, \rho^{l+1/2}\, L_n^{(l)}(a\rho^2)\, e^{-a\rho^2/2} \tag{5.51}$$

and are normalised as

$$\int_0^\infty \psi_n^{(l)}(\rho)^2\, d\rho = 1 \tag{5.52}$$

with

$$C_{n,l} = \sqrt{\frac{2n!}{(n+l)!}} a^{(l+1)/2},$$

where a is a structural parameter. Due to the bending nature of ρ and also the singularity at $\rho = \pi$, we will have to restrict the integration range in Eq. (5.52) to $\rho = [0, \ldots, \rho_{\max}]$, where $\rho_{\max} < \pi$ with $C_{n,l}$ obtained via numerical normalisation of $\psi_n^{(l)}(\rho)$.

The structural parameter a can be chosen as

$$a = \sqrt{\frac{f_2}{g_0}}, \tag{5.53}$$

where

$$f_2 = \frac{1}{2}\frac{\partial^2 \bar{V}(\rho)}{\partial \rho^2}\bigg|_{\rho=0}$$

and $g_0 = \bar{G}_{3,3}(\rho = 0)$ is the equilibrium value.

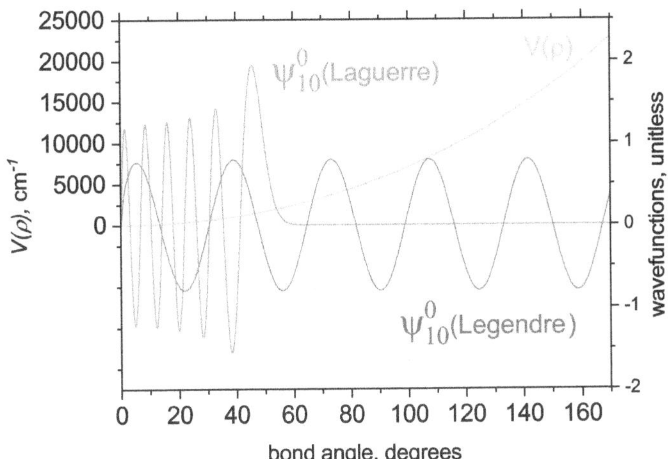

Figure 5.6 Example of a Legendre-polynomial-based basis function $\phi_{10}^{(0)}$ and a Laguerre-polynomial-based basis function $\psi_{10}^{(0)}$. The 1D potential energy function $V(\rho)$ associated with the latter is shown as reference.

With this choice of a, $\psi_n^{(l)}(\rho)$ in Eq. (5.51) are eigenfunctions of the model 1D Hamiltonian operator

$$\hat{H}^{\text{model}} = -\frac{\hbar^2}{2} g_0 \frac{\partial^2}{\partial \rho^2} + \frac{\hbar^2}{8} g_0 \frac{(4l^2 - 1)}{\rho^2} + f_2 \rho^2. \tag{5.54}$$

This 1D Hamiltonian operator is equivalent to that of the 2D IHO.

Figure 5.6 shows specific examples of a Legendre and a Laguerre basis functions, over-plotted with the potential energy function of CO_2 along the bending ρ coordinate: the Legendre basis function $\psi_{10}^{(k=0)}(\rho)$ is from Eq. (5.45), and the Laguerre $\psi_{10}^{(l=0)}(\rho)$ basis function is from Eq. (5.51). The Laguerre basis function has a much more compact form than the Legendre basis function and is therefore expected to provide a better bending description for the variational eigen-solution for semi-rigid systems as CO_2.

It should be noted that Legendre polynomials have better asymptotic behaviour at linear geometry ($\rho \to 0$) due to the factor $\sqrt{\sin\rho}$. Laguerre's normalisation term $\sqrt{\rho}$ provides only an approximate resolution of the singularity, which is, however, good enough for practical numerical applications. Even though the term $\sqrt{\rho}$ does not provide an exact resolution of the singularity as in the case of Legendre's $\sqrt{\sin\rho}$, one can see that the singularity disappears also in this case.

For example, by expanding the KEO around $\rho = 0$ and combined with the Laguerre basis functions $\psi_n^{(l)}$ for the main three singular KEO components, we obtain

$$\frac{\hbar^2}{2}\frac{\partial \psi_n^{(l)}}{\partial \rho} G_{3,3} \frac{\partial \psi_{n'}^{(l)}}{\partial \rho} + \psi_n^{(l)} U \psi_{n'}^{(l)} + \frac{\hbar^2}{2} G_{z,z} k^2 \psi_n^{(l)} \psi_{n'}^{(l)} =$$
$$\frac{\hbar^2}{2}\rho \frac{\partial \phi_n^{(l)}}{\partial \rho} G_{3,3} \frac{\partial \phi_{n'}^{(l)}}{\partial \rho} + \phi_n^{(l)} \tilde{U} \phi_{n'}^{(l)} + \frac{\hbar^2}{2} \rho\, \tilde{G}_{z,z} k^2 \chi_n^{(l)} \chi_{n'}^{(l)}, \quad (5.55)$$

where the following functions were introduced:

$$\phi_n^{(l)}(\rho) = C_{n,l}\, \rho^l\, L_n^{(l)}(a\rho^2)\, e^{-a\rho^2/2} \quad (l \geq 0), \quad (5.56)$$

$$\chi_n^{(l)}(\rho) = C_{n,l}\, \rho^{l-1}\, L_n^{(l)}(a\rho^2)\, e^{-a\rho^2/2} \quad (l > 0), \quad (5.57)$$

related to $\psi_n^{(l)}$ as

$$\psi_n^{(l)}(\rho) = \sqrt{\rho}\, \phi_n^{(l)}(\rho) = \sqrt{\rho}\, \rho\, \chi_n^{(l)}(\rho).$$

In Eq. (5.55)

$$\tilde{G}_{z,z} = G_{z,z}\rho^2,$$

$$\tilde{U} = U\rho + \frac{\hbar^2}{8}\left[-2\frac{\partial G_{3,3}}{\partial \rho} + \frac{G_{3,3}}{\rho}\right]$$

and we used the following property of the rotational basis function $|J,k,m\rangle$:

$$\hat{J}_z^2 G_{z,z}|J,k,m\rangle = \hbar^2 k^2 G_{z,z}|J,k,m\rangle.$$

It is easy to see from Eqs. (4.16, 4.23) that Eq. (5.55) is not singular. Indeed, the terms $\tilde{G}_{z,z}$ and $\tilde{U}$ can be Taylor expanded around $\rho = 0$ as follows:

$$\tilde{G}_{z,z} \approx \frac{1}{\mu_{XY}}\left[\frac{1}{r_1^2} + \frac{1}{r_2^2}\right] + \frac{2}{r_1 r_2 m_X} + O(\rho^2), \quad (5.58)$$

$$\tilde{U} \approx \hbar^2\left[-\frac{1}{6\mu_{XY}}\left(\frac{1}{r_1} + \frac{1}{r_2}\right) + \frac{2}{3 r_1 r_2 m_X}\right]\rho + O(\rho^3). \quad (5.59)$$

The term $G_{z,z}k^2\psi_n^{(l)}\psi_{n'}^{(l)}$ in Eq. (5.55) is either zero ($k = l = 0$) or finite if we assume that $l = |k|$.

The singularity of the $G_{x,x}$ term is even less of an issue for the Legendre basis functions due to the exponential term in Eq. (5.51), which ensures that the basis function vanishes at $\rho_{\max} < \pi$.

The Laguerre basis functions can be further optimised by solving a 1D bending Schrödinger equation for the model Hamiltonian operator

$$\hat{H}^{1D}(\rho) = -\frac{\hbar^2}{2}\frac{\partial}{\partial \rho}\bar{G}_{3,3}(\rho)\frac{\partial}{\partial \rho} + \frac{\hbar^2}{2}\bar{G}_{z,z}(\rho)k^2 + \bar{U}(\rho) + \bar{V}(\rho) \quad (5.60)$$

variationally on the basis $\psi_n^{(l)}(\rho)$. Here $\bar{G}_{3,3}(\rho)$, $\bar{G}_{z,z}(\rho)$, $\bar{U}$ and $\bar{V}$ are obtained from the corresponding 3D forms by setting $r_1 = r_2 = r_e$ (r_e is the equilibrium bond length).

5.6 ROTATIONAL BASIS SET

5.6.1 Rigid rotor wavefunctions

The choice of the rotational basis set is rather obvious and dictated by the conservation of the total angular momentum of the system in the isotropic environment, which we also assume here. In the absence of electronic and spin angular momenta, the total angular momentum $\boldsymbol{J}$ is identical to the rotational angular momentum $\boldsymbol{N}$. For this reason, in the following we will refer to the angular momentum as $\boldsymbol{J}$ and also use $\hat{J}$ as the corresponding operator. In the LF system, $\boldsymbol{J}$ has the three components J_X, J_Y and J_Z. J_x, J_y and J_z are the corresponding components in the molecular frame. The KEO in Eq. (3.5) does not contain the Euler angles ϕ, θ and χ explicitly, only via the operators $\hat{J}_x$, $\hat{J}_y$ and $\hat{J}_z$ (see Eqs. (2.14–2.16)), which is an intrinsic and important property of the Euler angles' representation.

The angular moments $\hat{J}_\alpha$ ($\alpha = x, y, z$) satisfy the anomalous commutation properties

$$[\hat{J}_\alpha, \hat{J}_\beta] = -i\hbar \hat{J}_\gamma,$$

where α, β and γ are x, y, z or their cyclic permutations and the angular momentum operator

$$\hat{J}^2 = \hat{J}_x^2 + \hat{J}_y^2 + \hat{J}_z^2$$

commutes with any $\hat{J}_\alpha$ components as well as with the Hamiltonian operator $\hat{H}$.

The rotational basis functions $\Phi_{\rm rot}$ are commonly from rigid rotor wavefunctions, i.e. eigenfunctions of $\hat{J}^2$, $\hat{J}_z$ and $\hat{J}_Z$:

$$\hat{J}^2 \Phi_{\rm rot} = J(J+1)\,\hbar^2\,\Phi_{\rm rot}, \tag{5.61}$$

$$\hat{J}_Z \Phi_{\rm rot} = m\,\hbar\,\Phi_{\rm rot}, \tag{5.62}$$

$$\hat{J}_z \Phi_{\rm rot} = k\,\hbar\,\Phi_{\rm rot}, \tag{5.63}$$

where J is the total angular momentum in units of $\hbar$, while m and k are the rotational quantum numbers, describing its projection onto laboratory-fixed Z and molecular z-axes, respectively, which satisfy

$$m, k = -J, -J-1, \ldots, 0, \ldots, J-1, J.$$

The rotational eigenfunctions satisfying these equation can be given by

$$|Jkm\rangle = \sqrt{\frac{2J+1}{8\pi^2}}[D^{(J)}_{mk}(\phi,\theta,\chi)]^*, \tag{5.64}$$

where $D^{(J)}_{mk}(\phi,\theta,\chi)$ is the known Wigner D-function satisfying

$$D^{(J)}_{mk}(\phi,\theta,\chi) = (-1)^{m-k}[D^{(J)}_{-m-k}(\phi,\theta,\chi)]^*.$$

In principle, J can be a half-integer in the presence of a half-integer total electron spin. For the purpose of this book, however, where only spin-free systems are considered, the total angular momentum quantum number J is an integer. The explicit definition of $D^{(J)}_{mk}(\phi,\theta,\chi)$ is very useful, and we reproduce it from Zare (1988) here:

$$D^{(J)}_{mk}(\phi,\theta,\chi) = \sqrt{(J+m)!(J-m)!(J+k)!(J-k)!}\, e^{-im\phi}e^{-ik\chi} \times \left[\sum_s (-1)^s \frac{(\cos\theta/2)^{2J+k-m-2s}(-\sin\theta/2)^{m-k+2s}}{s!(J-m-s)!(m-k+s)!(J+k-s)!}\right], \quad (5.65)$$

where the index s ranges over all integer values for which the factorial arguments are non-negative.

The matrix elements in this representation, in conjunction with the spherical harmonics, are zero between states of different J, which allows us to describe and process different values of the angular momentum J independently. This makes sense as J (the total angular momentum) is conserved. This property makes a huge impact on the ro-vibrational calculations. Even if it is not possible to fully separate the rotational and vibrational degrees of freedom, at least the ro-vibrational problem (i.e. ro-vibrational Hamiltonian matrix) is separable into subproblems (submatrices) of different J values. Other conserved properties, the so-called integrals of motion, include parity and, more generally, irreducible representations, discussed below, which can be used to further reduce the size of the Hamiltonian matrices.

5.6.2 Rotational matrix elements

For a spin-free molecule belonging to a Σ electronic state (in a field-free space), the rotational matrix elements of the KEO are between two rigid rotor wavefunctions in Eq. (5.64). The rotational dependence of the Hamiltonian operator is via the angular momenta operators $\hat{J}_\alpha$, and $\hat{J}_\alpha\hat{J}_\beta$, $\alpha,\beta = x,y,z$, and therefore only the matrix elements $\langle Jkm|\hat{J}_\alpha|Jk'm\rangle$ and $\langle Jkm|\hat{J}_\alpha\hat{J}_\beta|Jk'm\rangle$ are required. These matrix elements are diagonal both in m and J. Indeed, there is no dependence on m (projection of the angular momentum on the lab-axis Z) because all orientations in isotropic space are equivalent. The total angular momentum conserves, and there is no coupling between states of different J quantum numbers.

Non-zero rotational matrix elements of the angular momenta (linear and quadratic) are given by the standard relations:

$$\langle J,k,m|\hat{J}_z|J',k',m'\rangle = k\,\hbar\,\delta_{JJ'}\delta_{kk'}\delta_{mm'}, \quad (5.66)$$

$$\langle J,k\mp 1,m|\hat{J}_x|J,k,m\rangle = \frac{\hbar}{2}\,\left[(J\pm k)(J\mp k+1)\right]^{1/2}, \quad (5.67)$$

$$\langle J,k\mp 1,m|\hat{J}_y|J,k,m\rangle = \frac{\mp i\hbar}{2}\,\left[(J\pm k)(J\mp k+1)\right]^{1/2}, \quad (5.68)$$

$$\langle J,k\mp 1,m|\hat{J}_x\hat{J}_z|J,k,m\rangle = \frac{\hbar}{2}k\,\left[(J\pm k)(J\mp k+1)\right]^{1/2}, \quad (5.69)$$

$$\langle J,k\mp 1,m|\hat{J}_y\hat{J}_z|J,k,m\rangle = \mp\frac{i\hbar}{2}k\,\left[(J\pm k)(J\mp k+1)\right]^{1/2} \quad (5.70)$$

and

$$\langle J,k,m|\hat{J}^2|J,k,m\rangle = J(J+1)\,\hbar^2, \tag{5.71}$$

$$\langle J,k,m|\hat{J}_z^2|J,k,m\rangle = k^2\,\hbar^2, \tag{5.72}$$

$$\langle J,k,m|\hat{J}_x^2|J,k,m\rangle = \langle J,k,m|\hat{J}_y^2|J,k,m\rangle = \frac{\hbar^2}{2}\left[J(J+1)-k^2\right], \tag{5.73}$$

$$\langle J,k\mp 2,m|\hat{J}_x^2|J,k,m\rangle = -\langle J,k\mp 2,m|\hat{J}_y^2|J,k,m\rangle = \tag{5.74}$$

$$= \pm i\,\langle J,k\mp 2,m|\hat{J}_x\hat{J}_y|J,k,m\rangle = \tag{5.75}$$

$$= \frac{\hbar^2}{2}\left[J(J+1)-k(k\mp 1)\right]^{1/2}\left[J(J+1)-(k\mp 1)(k\mp 2)\right]^{1/2}, \tag{5.76}$$

where we used the phase choice of the rotational functions $|J,k,m\rangle$ as in Zare (1988). Introducing the standard ladder operators

$$\hat{J}_\pm = \hat{J}_x \pm \hat{J}_y, \tag{5.77}$$

for the corresponding matrix elements (non-vanishing only), one obtains

$$\langle J,k\mp 1,m|\hat{J}_\pm|J,k,m\rangle = \hbar\left[J(J+1)-k(k\mp 1)\right]^{1/2}, \tag{5.78}$$

$$\begin{aligned}\langle J,k-2,m|\hat{J}_+^2|J,k,m\rangle =& \hbar^2\left[J(J+1)-(k-1)(k-2)\right]^{1/2}\times\\ & \left[J(J+1)-k(k-1)\right]^{1/2},\end{aligned} \tag{5.79}$$

$$\begin{aligned}\langle J,k+2,m|\hat{J}_-^2|J,k,m\rangle =& \hbar^2\left[J(J+1)-(k+1)(k+2)\right]^{1/2}\\ & \left[J(J+1)-k(k+1)\right]^{1/2},\end{aligned} \tag{5.80}$$

$$\langle J,k,m|\hat{J}_\pm\hat{J}_\mp|J,k,m\rangle = \hbar^2\left[J(J+1)-k(k\pm 1)\right]. \tag{5.81}$$

It is essential to consider the representation of the so-called Wang functions

$$|J,K,\pm,m\rangle = \frac{1}{\sqrt{2}}\left[|J,K,m\rangle \pm |J,-K,m\rangle\right], \quad (K>0) \tag{5.82}$$

where the notation $K=|k|$ is introduced. One can show that with the right choice of the phases ± 1 in Eq. (5.82), all matrix elements of the KEO in Eq. (3.5) in the representation of the Wang functions can be made real. For example, the following Wang combinations with the phase $(-1)^{J+K+\tau}$ lead to real-valued matrix elements of the ro-vibrational Hamiltonian in Eq. (3.5):

$$|J,K,\tau\rangle = \frac{i^\tau}{\sqrt{2}}\left[|J,K,m\rangle + (-1)^{J+K+\tau}|J,-K,m\rangle\right], \tag{5.83}$$

$$|J,0,\tau\rangle = |J,0\rangle, \tag{5.84}$$

where $K>0$ and m is omitted for simplicity. Here the parameter $\tau=0,1$ defines the 'parity' of the Wang function $|J,K,\tau\rangle$ as follows:

$$E^*|J,K,\tau\rangle = (-1)^\tau|J,K,\tau\rangle,$$

where E^* is the inversion operation which, when applied to a molecule, inverts the spatial coordinates of all the nuclei and electrons through the molecular centre-of-mass. For $K = 0$ in Eq. (5.84), $J + \tau$ is always even. More specifically, for the Wang functions $|J, K, \tau\rangle$ defined in Eq. (5.83) with the phase choice $J + k + \tau$, the corresponding matrix elements of $i\hbar\hat{J}_x$, $i\hbar\hat{J}_y$ and $i\hbar\hat{J}_z$ are real.

Consider, for example, the Coriolis part of the KEO

$$\hat{T}^{\text{Cor}} = -i\hbar \sum_{\lambda,\alpha} \left[\frac{\partial}{\partial \xi_\lambda} G_{\lambda,\alpha} + G_{\alpha,\lambda} \frac{\partial}{\partial \xi_\lambda} \right] \hat{J}_\alpha.$$

Because of the imaginary factor $-i\hbar$, the integration of $\hat{J}_\alpha$ over the Wang functions in Eqs. (5.83, 5.84) produces real rotational matrix elements, with the non-zero components given by

$$\begin{aligned}
&\langle J, K, \tau | \hat{J}_x | J, K', \tau' \rangle \delta_{K,K\pm1} \\
&\langle J, K, \tau | \hat{J}_y | J, K', \tau \rangle \delta_{K,K\pm1} \\
&\langle J, K, \tau | \hat{J}_z | J, K', \tau' \rangle \delta_{K,K'},
\end{aligned}$$

where $\tau \neq \tau'$.

Consider, for example, matrix elements of $\langle J, K, \tau | \hat{J}_\alpha \hat{J}_\beta | J, K', \tau' \rangle$ from the pure rotational part of the KEO

$$\hat{T}^{\text{rot}} = \sum_{\alpha\beta} G_{\alpha\beta} \hat{J}_\alpha \hat{J}_\beta.$$

Using the parity-adapted Wang form in Eq. (5.83), all matrix elements are obtained as real expressions:

$$\begin{aligned}
\langle J, K, \tau | \hat{T}^{\text{rot}} | J, K', \tau \rangle = {} & \frac{1}{2} (G_{xx} - G_{yy}) \langle J, K, \tau | \left(\hat{J}_x^2 + \hat{J}_y^2 \right) | J, K', \tau \rangle \\
& + G_{x,z} \langle J, K, \tau | \left(\hat{J}_x \hat{J}_z + \hat{J}_z \hat{J}_x \right) | J, K', \tau \rangle \\
& + \left[\frac{1}{2} (G_{x,x} + G_{y,y}) \left(J(J+1) - K^2 \right) + G_{z,z} k^2 \right] \delta_{K,K'}, \quad (5.85)
\end{aligned}$$

$$\begin{aligned}
\langle J, K, \tau | \hat{T}^{\text{rot}} | J, K', \tau' \rangle = {} & G_{x,y} \langle J, K, \tau | \left(\hat{J}_x \hat{J}_y - \hat{J}_y \hat{J}_x \right) | J, K', \tau' \rangle \quad (5.86) \\
& + G_{y,z} \langle J, K, \tau | \left(\hat{J}_y \hat{J}_z + \hat{J}_z \hat{J}_y \right) | J, K', \tau' \rangle, \quad (5.87)
\end{aligned}$$

where $\tau' \neq \tau$ and the primitive rotational matrix elements are from Eqs. (5.69–5.74).

The vibrational wavefunctions can also be constructed to provide real matrix elements, if necessary.

MATERIALS USED

Description of orthogonal polynomials can be found in Abramowitz and Stegun (1964). Vibrational assignment by projecting on harmonic wavefunctions is discussed in Mátyus et al. (2010).

Calculations used for the example of the PH_3 polyad structure in Fig. 5.2 are from Sousa-Silva et al. (2013).

See Press et al. (2007) for numerical implementations of quadratures.

Double degenerate vibrational wavefunctions are from Bohaček et al. (1976) and Papoušek and Aliev (1982). The description of the 2D IHOs is from Nielsen (1951) and Bunker and Jensen (1998). See Yurchenko et al. (2005a) for application of the 2D isotropic wavefunctions in NH_3 ro-vibrational calculations. 3D isotropic vibrational basis set functions are from Messiah (1961), Papoušek and Aliev (1982) and Hougen (2001)

The Numerov-Cooley approach is from Noumerov (1924) and Cooley (1961).

The $J = 0$ band centre correction was explored in Yurchenko et al. (2011).

An example of monomer vibrational basis sets used for dimers is in Tennyson et al. (2012).

See on the quantum numbers assignment via projection of (ro-)vibrational wavefunctions, using the rigid rotor decomposition method in Szidarovszky et al. (2012) and vibrational configuration interaction technique in Mathea et al. (2021).

For the eigenfunction symmetrisation method, see Lemus (2003).

The construction of the vibrational basis set based on the vibrational angular momentum $\hat{L}$ was explored in Chubb et al. (2018).

The contracted subspaces methodology was used for PH_3 in Sousa-Silva et al. (2015), NH_3 in Yurchenko et al. (2011) and CH_4 in Yurchenko and Tennyson (2014).

The CO_2 potential energy function used in Fig. 5.6 is from Huang et al. (2013).

The Wang functions are by Wang (1929).

The directional cosines are from Wilson et al. (1955).

The Wigner D matrices, their properties and rotational matrix elements are from Zare (1988), Varshalovich et al. (1988), Bunker and Jensen (1998) and Bernath (2015).

The description of the Euler rotations is from Bunker and Jensen (1998) and Bernath (2015).

The properties of the angular momenta are from Papoušek and Aliev (1982), Zare (1988) and Bunker and Jensen (1998).

PES of PH_3 is from Sousa-Silva et al. (2015).

The phase choice $(-1)^{J+K+\tau}$ was used in Yurchenko et al. (2005a).

OTHER USEFUL MATERIALS NOT COVERED IN THIS BOOK

Using the matrix-vector products (Lanczos) to solve the nuclear motion problems (Carrington, 2018).

DVR algorithms (Light et al., 1985; Bačić and Light, 1989; Bramley and Carrington, 1993; Light and Carrington Jr., 2000; Tennyson et al., 2004; Mátyus et al., 2009a).

Examples of systematic multi-layer contraction schemes include MCTDH by Wang (2015) and Sarka and Poirier (2021).

CHAPTER 6

Symmetry-adapted basis sets

In this chapter we introduce the aspect of molecular symmetry into our ro-vibrational treatment and show how a symmetry-adapted basis set can facilitate a variational solution. Some basic irreducible representation algebra of molecular symmetry groups is described. Computational procedures for symmetrisation of basis sets are introduced, including the eigenfunction method based on diagonalisation of reduced Hamiltonian matrices and the wavefunction-sampling method. Numerical examples are used to illustrate their step-by-step implementation. Finally we expand on symmetrisation of the rigid rotor functions that require special treatment.

6.1 INTRODUCTION

There are a number of reasons why a symmetrisation of the basis set is important. The most obvious one is to help reduce the size of the Hamiltonian matrix. Indeed, because of the Hamiltonian operator being fully symmetric, Hamiltonian matrix elements are automatically zero between states of different symmetries. This can be nicely illustrated by the following example of two basis functions of an XY_2 molecule (e.g. H_2S from the example in Section 5.4.5) given by

$$\phi^{\mathrm{s}}_{v_1,v_2}(r_1,r_2) = \frac{1}{\sqrt{2}}\left[\phi_{v_1}(r_1)\phi_{v_2}(r_2) + \phi_{v_2}(r_1)\phi_{v_1}(r_2)\right],$$
$$\phi^{\mathrm{a}}_{v_1,v_2}(r_1,r_2) = \frac{1}{\sqrt{2}}\left[\phi_{v_1}(r_1)\phi_{v_2}(r_2) - \phi_{v_2}(r_1)\phi_{v_1}(r_2)\right].$$

Here $\phi_{v_1}(r_1)$ and $\phi_{v_2}(r_2)$ are equivalent 1D stretching basis functions describing identical nuclei, which we combine into symmetric and antisymmetric combinations. It is easy to see that a permutation of the indexes v_1 and v_2 only affects their phases: $\phi^{\mathrm{s}}_{v_1,v_2}$ (symmetric) is unchanged, while the sign of $\phi^{\mathrm{a}}_{v_1,v_2}$ has changed its sign. The same effect is observed if the nuclear coordinates are swapped:

$$(12)\phi^{\mathrm{s}}_{v_1,v_2}(r_1,r_2) = \phi^{\mathrm{s}}_{v_1,v_2}(r_2,r_1) = \phi^{\mathrm{s}}_{v_1,v_2}(r_1,r_2),$$
$$(12)\phi^{\mathrm{a}}_{v_1,v_2}(r_1,r_2) = \phi^{\mathrm{a}}_{v_1,v_2}(r_2,r_1) = -\phi^{\mathrm{a}}_{v_1,v_2}(r_1,r_2),$$

DOI: 10.1201/9780429154348-6

where the notation (12) for the operation $1 \leftrightarrow 2$ was introduced. The effect of the permutation $1 \leftrightarrow 2$ is to interchange nuclei 1 and 2 as illustrated in Fig. 6.1. The Hamiltonian operator $\hat{H}(r_1, r_2)$ is invariant to any symmetry operation in the sense that any permutation (and inversion) is to leave $\hat{H}$ unchanged (see Bunker and Jensen (1998) for details), including (12)

$$(12)\hat{H} = \hat{H}.$$

Let us consider the Hamiltonian matrix elements between these wavefunctions:

$$\langle \phi^{\mathrm{s}}_{v_1,v_2} | \hat{H}(r_1, r_2) | \phi^{\mathrm{a}}_{v_1',v_2'} \rangle = \int \phi^{\mathrm{s}*}_{v_1,v_2} \hat{H}(r_1, r_2) \phi^{\mathrm{a}}_{v_1',v_2'} d\tau,$$

where we interchange the integration variables r_1 and r_2:

$$\int \phi^{\mathrm{s}*}_{v_1,v_2} \hat{H}(r_1, r_2) \phi^{\mathrm{a}}_{v_1',v_2'} d\tau = \int \phi^{\mathrm{s}*}_{v_1,v_2}(r_2, r_1) \hat{H}(r_2, r_1) \phi^{\mathrm{a}}_{v_1',v_2'}(r_2, r_1) d\tau.$$

Using their permutation properties, we obtain that the matrix element must vanish to satisfy

$$\langle \phi^{\mathrm{s}}_{v_1,v_2} | \hat{H}(r_1, r_2) | \phi^{\mathrm{a}}_{v_1',v_2'} \rangle = -\langle \phi^{\mathrm{s}}_{v_1,v_2} | \hat{H}(r_1, r_2) | \phi^{\mathrm{a}}_{v_1',v_2'} \rangle = 0. \tag{6.1}$$

Here we assumed that the integration volume $d\tau$ is unchanged upon the permutation (12). This is a general property that Hamiltonian operator is block-diagonal in the representation of symmetry-adapted basis functions.

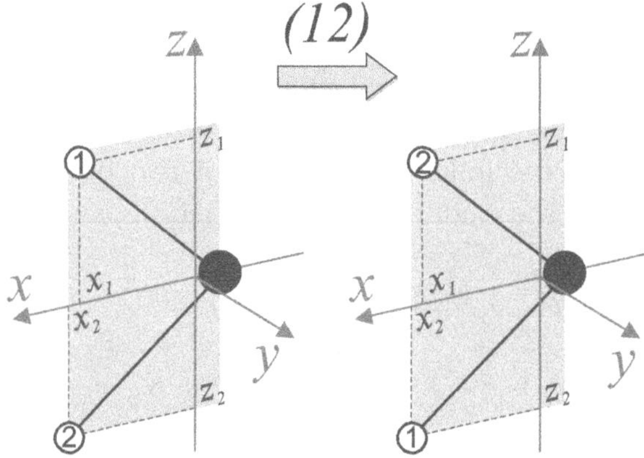

Figure 6.1 The permutation operation (12) on the coordinates r_1 and r_2 of an XY_2 molecule.

Another, even more important reason is that the symmetry specification of the ro-vibrational wavefunctions is related to the nuclear spins, which we have ignored so far. The nuclear statistics play an important role in intensity calculations and are affected by the symmetry of the ro-vibrational wavefunction via the Pauli principle, which will be discussed in Section 7.1.1 below.

In addition to the permutation operator, let us also introduce an inversion operation E^*, which is defined as the operation of inverting the spatial coordinates of all the nuclei and electrons through the molecular centre-of-mass, as illustrated in Fig. 6.2 for an XY_2 molecule. As one can see, E^* does not affect the values of the valence coordinates r_1 and r_2 and, therefore, does not affect $\phi^{s}_{v_1,v_2}$ and $\phi^{a}_{v_1,v_2}$.

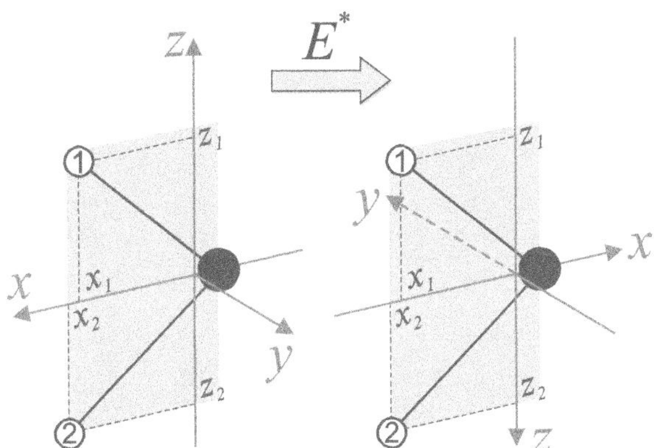

Figure 6.2 The inversion operation E^* on the coordinates r_1 and r_2 of an XY_2 molecule.

In this book we follow the concept of the molecular symmetry (MS) group (Bunker and Jensen, 1998) to classify the ro-vibrational levels and states of a molecule. The MS group originated in the works of Longuet-Higgins and Hougen which were based on the permutation and inversion properties of identical nuclei and the corresponding invariance of the molecular Hamiltonian. It is conventionally indicated by appending '(M)' to the group symbol, e.g. $\mathcal{C}_{2v}$(M). One of the important aspects of molecular symmetry groups is that they deal with the degrees of freedom associated with moving identical particles as opposed to the molecular point groups, which operate on the positions of identical atoms in their static (equilibrium) configuration. For semi-rigid molecules, molecular symmetry groups are equivalent (isomorphic) to the corresponding point groups and can be interchangeably used by mapping their group operations to each other. In Table 6.1 equivalent symmetry operations of the point group $\mathcal{C}_{2v}$ and molecular symmetry group $\mathcal{C}_{2v}$(M) are listed. The situation is different for non-rigid systems, where molecular symmetry groups, with their concept of feasible operations, contain group elements that do not have immediate analogy in the corresponding point groups of the same molecule. One of the famous examples is the molecule NH_3, with the equilibrium structure described by the point group symmetry $\mathcal{C}_{3v}$, whose rotation-inversion-vibrational motion requires the molecular symmetry group $\mathcal{D}_{3h}$(M). For a comprehensive description of this subject, see the book 'Molecular symmetry' by Bunker and Jensen (1998).

A symmetry-adapted basis set is constructed as a linear combination of basis functions that transform according to one of the irreducible representations (irreps) Γ of the molecular symmetry group in question. We thus aim to build and use

Table 6.1 The transformation of the internal coordinates of an XY_3-type molecule by the symmetry operations of the molecular symmetry group $\mathcal{C}_{3v}$(M) and the corresponding equivalent operations of the point group $\mathcal{C}_{3v}$. (M) indicates the molecular symmetry group operators, while PG stands for the point group.

Variables	(M)	E	(123)	(321)	(23)*	(13)*	(12)*
	PG	E	C_3	$2C_3$	σ_v	σ_v	σ_v
r_1		r_1	r_3	r_2	r_1	r_3	r_2
r_2		r_2	r_1	r_3	r_3	r_2	r_1
r_3		r_3	r_2	r_1	r_2	r_1	r_3
α_{23}		α_{23}	α_{12}	α_{13}	α_{23}	α_{12}	α_{13}
α_{13}		α_{13}	α_{23}	α_{12}	α_{12}	α_{13}	α_{23}
α_{12}		α_{12}	α_{13}	α_{23}	α_{13}	α_{23}	α_{12}

symmetry-adapted ro-vibrational basis functions $\Phi_i^{J,\Gamma}$ that transform as an irrep Γ (or 'symmetry') of the group $\boldsymbol{G}$ the molecule belongs to. Here i is a counting number.

The corresponding Hamiltonian matrix in the representation $\Phi_i^{J,\Gamma}$ has a block-diagonal form (compare to Eq. (6.1)):

$$\langle \Phi_i^{J,\Gamma} | H^{\rm rv} | \Psi_{i'}^{J',\Gamma'} \rangle = H_{i,i'} \delta_{\Gamma,\Gamma'} \delta_{J,J'}. \tag{6.2}$$

In practice, this means that each (J,Γ)-block can be diagonalised independently with J and Γ being good quantum 'numbers' (i.e. constants of motion).

Consider, for example, the rotation-vibrational motion of a rigid molecule PH_3 in some instantaneous configuration described by the six valence coordinates P-$H_1 = r_1$, P-$H_2 = r_2$, P-$H_3 = r_3$, H_1-P-$H_2 = \alpha_3$, H_1-P-$H_3 = \alpha_2$ and H_2-P-$H_3 = \alpha_1$ as illustrated in Fig. 6.3. The point group symmetry of PH_3 is $\mathcal{C}_{3v}$, while the nuclear degrees of freedom span the molecular symmetry group $\mathcal{C}_{3v}$(M). Both groups consist of three irreps $\Gamma = A_1$, A_2 and E, with the group operations R and corresponding characters listed in Tables 6.1 and 6.2 (see below for the definition of the characters). The E-symmetry irrep is two-fold, with two degenerate components E_a and E_b.

The elements of $\mathcal{C}_{3v}$(M) are operations on the (motion of the) identical nuclei H_i: cycling permutations (123) and (321) and permutations with inversion (23)*, (13)* and (12)*. The operation (23)* is an interchange $2 \leftrightarrow 3$ with inversion, which is a feasible operation, in contrast to (23), which is unfeasible operation (i.e. separated by insuperable barrier). We follow the notation of the permutation operation $(abcd\ldots)$ of Bunker and Jensen (1998), which is illustrated in Fig. 6.3 for the operation (123): (123) replaces nucleus 1 by 2, so that nucleus 2 has its new coordinates those which nucleus 1 had. Also nucleus 2 is replaced by 3 and nucleus 3 is replaced by 1, inheriting the corresponding coordinates. Thus, the bond P-H_1 has now length r_3, the bond P-H_2 has length r_1 and bond P-H_3 has length r_2, i.e. for

Table 6.2 Characters of the molecular symmetry group $\mathcal{C}_{3v}$(M).

	Characters		
Irrep Γ	E	(123) (321)	(23)* (13)* (12)*
A_1	1	1	1
A_2	1	1	-1
E	2	-1	0

(123), $r_1 \to r_3$, $r_2 \to r_1$ and $r_3 \to r_2$. Accordingly, the new inter-bond angles are H_1-P-H_2 $= \alpha_2$, H_1-P-H_3 $= \alpha_1$ and H_2-P-H_3 $= \alpha_3$, i.e. $\alpha_1 \to \alpha_3$, $\alpha_2 \to \alpha_1$ and $\alpha_3 \to \alpha_2$.

Let us now consider stretching primitive basis functions in the product form

$$\phi_{v_1,v_2,v_3}(r_1,r_2,r_3) = |v_1\rangle|v_2\rangle|v_3\rangle \tag{6.3}$$

and use them to construct linear combinations Φ_i^Γ ($\Gamma = A_1$, A_2 and E) that transform in accordance with the corresponding irreps Γ. This means that when applying a group operation $R \in \boldsymbol{G}$ to an l-fold degenerate wavefunction $\boldsymbol{\Phi}_i^\Gamma = (\Phi_{i,1}^\Gamma, \Phi_{i,2}^\Gamma, \ldots, \Phi_{i,l}^\Gamma)^T$, the result is a unitary transformation

$$R\boldsymbol{\Phi}_i^\Gamma = \boldsymbol{D}^\Gamma[R]\boldsymbol{\Phi}_i^\Gamma,$$

where $\boldsymbol{D}^\Gamma[R]$ is an $l \times l$ matrix with the sum of diagonal elements called the character χ of Γ for the operation R (see Section 6.2 for more details on the group characters). Characters are used to identify group irreps as, e.g., in the case of $\mathcal{C}_{3v}$(M) in Table 6.2, where the representations A_1, A_2 and E are uniquely identified by the combinations $\{1,1,1\}$, $\{1,1,-1\}$ and $\{2,-1,0\}$, respectively.

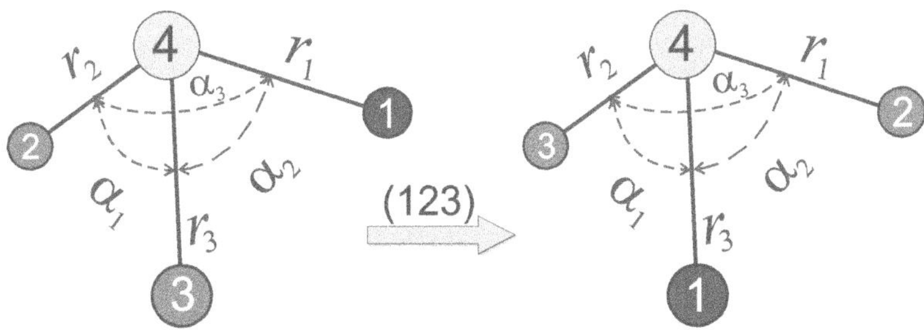

Figure 6.3 The operation of (123) on the coordinates r_1, r_2, r_3, α_1, α_2 and α_3 of a rigid PH_3 molecule.

It is easy to see that, for example, a ground state combination $|0\rangle|0\rangle|0\rangle$ is unchanged upon any group operation and is therefore fully symmetric. Indeed, for the permutation operation (123), the transformation is given by

$$(123)\,\phi_0(r_1)\phi_0(r_2)\phi_0(r_3) = \phi_0(r_3)\phi_0(r_1)\phi_0(r_2),$$

which is identical to $\phi_0(r_1)\phi_0(r_2)\phi_0(r_3)$. Therefore, the transformational matrix $\boldsymbol{D}^{\Gamma}[(123)]$ in this case is single valued, the only element of $D^{\Gamma}_{1,1}[(123)]$, which gives the character of $\chi[(123)] = 1$. Analogously, group characters of all other transformations of $|0\rangle|0\rangle|0\rangle$ are 1, and therefore, according to Table 6.2, $\phi_0(r_1)\phi_0(r_2)\phi_0(r_3)$ have the symmetry $\Gamma = A_1$.

One can show (see below) that the following combinations of primitive basis functions $\phi_{v_1,v_2,v_3}(r_1,r_2,r_3) \equiv |v_1,v_2,v_3\rangle$ transform correctly as one of the irreducible representations of $\mathcal{C}_{3\mathrm{v}}(\mathrm{M})$

$$|vvv;A_1\rangle = |v,v,v\rangle; \tag{6.4}$$

$$|vvv';A_1\rangle = \frac{1}{\sqrt{3}}\left(|v,v,v'\rangle + |v,v',v\rangle + |v',v,v\rangle\right), \tag{6.5}$$

$$|vvv';E_a\rangle = \frac{1}{\sqrt{6}}\left(2|v,v,v'\rangle - |v,v',v\rangle - |v',v,v\rangle\right), \tag{6.6}$$

$$|vvv';E_b\rangle = \frac{1}{\sqrt{2}}\left(|v',v,v\rangle - |v,v',v\rangle\right). \tag{6.7}$$

To illustrate the symmetry properties of these combinations, let us consider an example of the two-component wavefunction of symmetry E, $(|vvv';E_a\rangle, |vvv';E_b\rangle)^T$ from Eqs. (6.6, 6.7). It can be shown that it transforms upon the operation (123) as given by

$$(123)\begin{pmatrix} |vvv';E_a\rangle \\ |vvv';E_b\rangle \end{pmatrix} = \begin{pmatrix} -\frac{1}{2} & \frac{\sqrt{3}}{2} \\ -\frac{\sqrt{3}}{2} & -\frac{1}{2} \end{pmatrix}\begin{pmatrix} |vvv';E_a\rangle \\ |vvv';E_b\rangle \end{pmatrix}. \tag{6.8}$$

The sum of the diagonal elements of

$$\boldsymbol{D}^{\Gamma}[(123)] = \begin{pmatrix} -\frac{1}{2} & \frac{\sqrt{3}}{2} \\ -\frac{\sqrt{3}}{2} & -\frac{1}{2} \end{pmatrix} \tag{6.9}$$

is -1, i.e. the same as the character of $R = (123)$ in Table 6.2, which points to the irrep (symmetry) $\Gamma = E$.

Our goal here is to describe a general (numerically motivated) procedure to build a symmetry-adapted set of ro-vibrational basis functions Φ^{Γ}_{μ} that transform as irreducible representations Γ of $\boldsymbol{G}$. In the case of degenerate, e.g. l-fold irreps, an additional subscript $n = 1,\ldots,l$ will be used to refer to a specific degenerate component, $\Phi^{\Gamma}_{\mu,n}$. For example, for the two-fold E symmetry functions $|vvv';E_a\rangle, |vvv';E_b\rangle$ in Eqs. (6.6, 6.7), $n = 1$ and $n = 2$ will be used to distinguish the E_a and E_b symmetry components, respectively.

A symmetry adaptation of the primitive ro-vibrational basis functions $\phi^{J}_{k,\nu}$ can be sought as a finite linear combination

$$\Phi^{J,\Gamma}_{\mu,n} = \sum_{k,\nu} T^{\mu,J,\Gamma}_{k,\nu,n} \phi^{J}_{k,\nu} \,, \tag{6.10}$$

where $T^{\mu,J,\Gamma}_{k,\nu,n}$ are symmetrisation coefficients. For the PH_3 primitive functions from Eq. (5.3), $T^{\mu,J,\Gamma}_{k,\nu,n}$ are the expansion coefficients in front of $|v_1, v_2, v_3\rangle$ in Eqs. (6.5–6.7). We will show how construction of symmetrisation coefficients $T^{\mu,J,\Gamma}_{k,\nu,n}$ can be represented using a general numerical procedure.

It should be noted that using symmetries in energy calculations is advantageous but not critical. Indeed, a variational solution with a non-symmetrised basis set should be as correct as with the symmetry-adapted basis, provided it is variationally accurate. However, symmetry is mandatory in intensity calculations because of the nuclear statistics (see the degeneracy factor $g_{\rm ns}$ in Eq. (7.16)), at least for hyperfine unresolved states, i.e. when the interaction of the ro-vibrational motion with the nuclear spins is neglected. The nuclear statistical factors are intrinsically built on the symmetric, permutation properties of nuclei (bosons or fermions) to satisfy the Pauli principle and cannot be used without knowing the state symmetry.

6.2 MOLECULAR SYMMETRY: SOME THEORY ON THE REDUCTION TO IRREDUCIBLE REPRESENTATIONS, PROJECTION TECHNIQUE AND SYMMETRY CLASSIFICATIONS

Let us consider a set of ro-vibrational (basis) functions ϕ_n, which we aim to symmetrise according to a symmetry group $\boldsymbol{G}$ by combining some 'equivalent' primitive functions. Let us assume that the result of group operations R on such equivalent functions ϕ_n can be represented by linear combinations involving these symmetry equivalent functions as follows:

$$R\,\phi_n = \sum_{n'} D^{\Gamma}_{n,n'}[R]\,\phi_{n'};$$

the set of matrices $\boldsymbol{D}^{\Gamma}[R]$ are said to form a representation Γ of the group $\boldsymbol{G}$. In general, Γ is a reducible representation, which means that the representation matrices $\boldsymbol{D}^{\Gamma}[R]$ can be unitarily transformed to block diagonal forms of smaller dimensions. The transformation procedure is called a reduction of Γ to a combination of irreducible representations Γ_s as given by

$$\Gamma = a_1\Gamma_1 \oplus a_2\Gamma_2 \oplus a_3\Gamma_3 \oplus \cdots \,; \tag{6.11}$$

s is introduced to count different irreps of the same symmetry, and a_s are reduction coefficients, i.e. corresponding numbers of irreducible representations of the same symmetry Γ_s.

The reduction coefficients can be obtained using the so-called group characters $\chi[R]$ of the reducible representation Γ defined as traces of the transformation matrices $D_{m,n}[R]$ as follows:

$$\chi[R] = \sum_n D_{n,n}[R].$$

The reduction coefficients a_s for each irrep $\Gamma_s \in \boldsymbol{G}$ are then given by

$$a_s = \frac{1}{g} \sum_{R \in \boldsymbol{G}} \chi[R] \, \chi^{\Gamma_s}[R]^*, \tag{6.12}$$

where g is the order of the group (number of elements), R runs over all the elements of the group, $\chi^{\Gamma_s}[R]$ are (known tabulated values) group characters and $*$ indicates complex conjugation. The coefficients a_s satisfy the reduction relations

$$\chi[R] = \sum_s a_s \chi^{\Gamma_s}[R] \quad \text{and} \quad \sum_R \left|\chi^{\Gamma_s}[R]\right|^2 = g. \tag{6.13}$$

For a $\mathcal{C}_{3\text{v}}$-type molecule like PH_3, the characters $\chi^{\Gamma_s}[R]$ are listed in Table 6.2.

The standard technique to construct irreducible representations (i.e. symmetry-adapted combinations) is by using projection operators (see, e.g., Bunker and Jensen (1998)). For a non-degenerate representation, a projection operator is trivially constructed as a linear combination of the group operations weighted by the corresponding group characters as given by

$$P^{\Gamma_s} = \frac{1}{g} \sum_{R \in \boldsymbol{G}} \chi^{\Gamma_s}[R]^* R. \tag{6.14}$$

A degenerate case requires special care. For an l_s-fold degenerate representation ($l_s > 1$), the projection operator is defined as

$$P^{\Gamma_s} = \frac{l_s}{g} \sum_{R \in \boldsymbol{G}} \chi^{\Gamma_s}[R]^* R. \tag{6.15}$$

In the reducible representation of the group, the projection operator is a matrix, given by a weighted sum of reducible matrix representations $\boldsymbol{D}^{\Gamma}[R]$ as follows:

$$\boldsymbol{P}^{\Gamma_s} = \frac{l_s}{g} \sum_{R \in \boldsymbol{G}} \chi^{\Gamma_s}[R]^* \boldsymbol{D}^{\Gamma}[R].$$

By applying $\boldsymbol{P}^{\Gamma_s}$ to non-symmetrised basis components ϕ_n, at least one unique irrep component $\Phi_m^{\Gamma_s}$ can be generated. Other degenerate components $\Phi_n^{\Gamma_s}$ of the same degenerate irrep can be recovered using the transfer operator

$$P_{mn}^{\Gamma_s} = \frac{l_s}{g} \sum_R D^{\Gamma_s}[R]_{mn}^* R, \quad m \neq n, \tag{6.16}$$

by acting it on $\Phi_m^{\Gamma_s}$,

$$P_{mn}^{\Gamma_s}\Phi_m^{\Gamma_s} = \Phi_n^{\Gamma_s},$$

where $\boldsymbol{D}^{\Gamma_s}[R]$ is an irreducible orthogonal transformation matrix of Γ_s for the operation R.

The intrinsic property of the degenerate wavefunctions is that their definition is not unique as any mixture of degenerate eigenfunctions of $\hat{H}$ is an eligible solution. However, it is practical to establish and follow some specific standards for the representation of the degenerate components. This helps transparency between different parts of the application and even between different applications and programs. For the two-dimensional symmetry E, it is common to impose the condition that the E symmetry components, E_a and E_b, transform as a spatial vector (x, y) upon some equivalent rotation. An equivalent rotation is an operation describing the change in the Euler angles upon a molecular group transformation. For example, in the case of $\mathcal{C}_{3\mathrm{v}}(\mathrm{M})$, the operation (123) is equivalent to the rotation of a unit vector by 120 degrees through the coordinate centre in a clockwise manner, i.e.

$$(123)\begin{pmatrix} E_a \\ E_b \end{pmatrix} = \begin{pmatrix} -\frac{1}{2} & \frac{\sqrt{3}}{2} \\ -\frac{\sqrt{3}}{2} & \frac{1}{2} \end{pmatrix}\begin{pmatrix} E_a \\ E_b \end{pmatrix}, \tag{6.17}$$

while the operation (12) is equivalent to the reflection of the vector (x, y) through the y axis:

$$(12)\begin{pmatrix} E_a \\ E_b \end{pmatrix} = \begin{pmatrix} 1 & 0 \\ 0 & -1 \end{pmatrix}\begin{pmatrix} E_a \\ E_b \end{pmatrix}. \tag{6.18}$$

The transformation matrices for all six group operations of $\mathcal{C}_{3\mathrm{v}}$ are listed in Table 6.4.

Let us use the example of the $\mathcal{C}_{3\mathrm{v}}(\mathrm{M})$ functions $|v_1, v_2, v_3\rangle$ of PH_3 and illustrate how (some of) their symmetry adaptations can be constructed using the projection technique. The function $|v, v, v\rangle$ ($v_1 = v_2 = v_3$) in Eq. (6.4) already transforms as A_1 upon any $\mathcal{C}_{3\mathrm{v}}(\mathrm{M})$ symmetry operations from Table 6.2 and hence does not need any further reduction.

Let us consider three equivalent combinations of the primitive basis functions $\phi_1 = |vvv'\rangle$, $\phi_2 = |vv'v\rangle$ and $\phi_3 = |v'vv\rangle$ and construct a reduction for them as in Eqs. (6.5–6.7). We start with a reconstruction of the reducible representation and associated group characters by applying the corresponding symmetry operations of the group. For example, by applying (123) to the vector (ϕ_1, ϕ_2, ϕ_3), we obtain

$$(123)\begin{pmatrix} \phi_1 \\ \phi_2 \\ \phi_3 \end{pmatrix} = \begin{pmatrix} 0 & 0 & 1 \\ 1 & 0 & 0 \\ 0 & 1 & 0 \end{pmatrix}\begin{pmatrix} \phi_1 \\ \phi_2 \\ \phi_3 \end{pmatrix}, \tag{6.19}$$

where the 3×3 transformation matrix has the character 0 (sum of the diagonal elements). All other reducible representation matrices can be obtained analogously and are listed in Table 6.3, together with their characters. By applying the reduction relation in Eq. (6.3) with $\chi[R]$ from Table 6.3 and $\chi^{\Gamma_s}[R]$ from Table 6.2 (the group

Table 6.3 Reducible $\mathcal{C}_{3\text{v}}$ transformation properties of the basis functions $\phi_1 = |vvv'\rangle$, $\phi_2 = |vv'v\rangle$ and $\phi_3 = |v'vv\rangle$ upon operations R.

R	Transformation	$\chi[R]$	R	Transformation	$\chi[R]$
E	$\begin{pmatrix} 1 & 0 & 0 \\ 0 & 1 & 0 \\ 0 & 0 & 1 \end{pmatrix}$	3	(23)*	$\begin{pmatrix} 1 & 0 & 0 \\ 0 & 0 & 1 \\ 0 & 1 & 0 \end{pmatrix}$	1
(123)	$\begin{pmatrix} 0 & 0 & 1 \\ 1 & 0 & 0 \\ 0 & 1 & 0 \end{pmatrix}$	0	(13)*	$\begin{pmatrix} 0 & 0 & 1 \\ 0 & 1 & 0 \\ 1 & 0 & 0 \end{pmatrix}$	1
(321)	$\begin{pmatrix} 0 & 1 & 0 \\ 0 & 0 & 1 \\ 1 & 0 & 0 \end{pmatrix}$	0	(12)*	$\begin{pmatrix} 0 & 1 & 0 \\ 1 & 0 & 0 \\ 0 & 0 & 1 \end{pmatrix}$	1

order of $\mathcal{C}_{3\text{v}}$(M) is 6), we obtain $a_{A_1} = 1$, $a_{A_2} = 0$ and $a_E = 1$. This appears to be consistent with the number of symmetry-adapted combinations in Eqs. (6.5–6.7).

In order to obtain the actual symmetry-adapted (irreducible) combinations of wavefunctions, we follow Eq. (6.14) and form projection operators. For A_1 ($a_{A_1} = 1$), we obtain

$$\boldsymbol{P}^{A_1} = \begin{pmatrix} \frac{1}{3} & \frac{1}{3} & \frac{1}{3} \\ \frac{1}{3} & \frac{1}{3} & \frac{1}{3} \\ \frac{1}{3} & \frac{1}{3} & \frac{1}{3} \end{pmatrix}.$$

By taking any of the identical rows, a non-normalised vector with the identical coefficients 1/3 is obtained:

$$\Psi^{A_1} = \frac{1}{3} \begin{pmatrix} 1 \\ 1 \\ 1 \end{pmatrix},$$

which, after normalisation coincides with Eq. (6.5).

An attempt to pull an irrep for $\Gamma_s = A_2$ from the reducible representation in Table 6.3 leads to a projection operator P^{A_2} that annihilates ϕ_1, ϕ_2 and ϕ_3, leading to the trivial solution of $\Psi^{A_2} = 0$. The E irrep is, however, not trivial and provides an instructive example. Following the irrep protocol, the $\Gamma_s = E$ projection operator is obtained using Eq. (6.15) in conjunction with the group characters $\chi^{\Gamma_s} = 2, -1, 0$ from Table 6.2, the reducible matrices $D^{\Gamma}[R]$ from Table 6.3 and assuming $g = 6$ and $l_s = 2$. The following matrix representation of $\boldsymbol{P}^E$ is then obtained

$$\boldsymbol{P}^E = \frac{1}{3} \begin{pmatrix} 2 & -1 & -1 \\ -1 & 2 & -1 \\ -1 & -1 & 2 \end{pmatrix}.$$

Let us pick the first row, which after normalisation gives the first E symmetry component

$$\Psi_1^E = \frac{1}{\sqrt{6}} \begin{pmatrix} 2 \\ -1 \\ -1 \end{pmatrix}.$$

Table 6.4 Irreducible $\mathcal{C}_{3\mathrm{v}}$ transformation matrices of the E symmetry upon operations R.

R	Transformation	$\chi[R]$	R	Transformation	$\chi[R]$
E	$\begin{pmatrix} 1 & 0 \\ 0 & 1 \end{pmatrix}$	2	$(23)^*$	$\begin{pmatrix} -\frac{1}{2} & \frac{\sqrt{3}}{2} \\ \frac{\sqrt{3}}{2} & \frac{1}{2} \end{pmatrix}$	0
(123)	$\begin{pmatrix} -\frac{1}{2} & \frac{\sqrt{3}}{2} \\ -\frac{\sqrt{3}}{2} & \frac{1}{2} \end{pmatrix}$	-1	$(13)^*$	$\begin{pmatrix} -\frac{1}{2} & -\frac{\sqrt{3}}{2} \\ -\frac{\sqrt{3}}{2} & \frac{1}{2} \end{pmatrix}$	0
(321)	$\begin{pmatrix} -\frac{1}{2} & -\frac{\sqrt{3}}{2} \\ \frac{\sqrt{3}}{2} & \frac{1}{2} \end{pmatrix}$	-1	$(12)^*$	$\begin{pmatrix} 1 & 0 \\ 0 & -1 \end{pmatrix}$	0

The second component is reconstructed with the help of the transfer operator P^E_{12} from Eq. (6.16) using the following elements $D^E_{12}[R]^*$ from Table 6.4

$$0, \sqrt{3}/2, -\sqrt{3}/2, \sqrt{3}/2, -\sqrt{3}/2 \quad \text{and} \quad 0$$

for $R = E$, (123), (321), $(23)^*$, $(13)^*$ and $(12)^*$, respectively. The transfer operator is thus given by

$$\boldsymbol{P}^E_{12} = \frac{\sqrt{3}}{6} \begin{pmatrix} 1 & -1 & 0 \\ 1 & -1 & 0 \\ -2 & 1 & 1 \end{pmatrix}.$$

Now, Ψ^E_2 is obtained by applying $\boldsymbol{P}^E_{12}$ to Ψ^E_1 and is after normalisation given by

$$\Psi^E_2 = \frac{1}{\sqrt{6}} \begin{pmatrix} 1 \\ 1 \\ -2 \end{pmatrix}.$$

Although Ψ^E_2 is independent and normalised, it is not orthogonal to Ψ^E_1:

$$\langle \Psi^E_1 | \Psi^E_2 \rangle = \frac{5}{6}.$$

Orthogonality is an important property of irreducible representations and must be respected. It is conveniently achieved via the Gram-Schmidt technique:

$$\Psi'^E_2 = |\Psi^E_2\rangle - \langle \Psi^E_1 | \Psi^E_2 \rangle |\Psi^E_1\rangle,$$

so our second component now reads

$$\Psi'^E_2 = \frac{\sqrt{3}}{2\sqrt{2}} \begin{pmatrix} 0 \\ 1 \\ -1 \end{pmatrix},$$

which after the final normalisation is found to coincide with the corresponding expressions in Eqs. (6.6, 6.7). Another numerical example of symmetrisation of the E-symmetry components will be provided below, in Section 6.5.1.

As part of the basis set contraction, in the following we present a robust symmetrisation technique based on diagonalisation of reduced Hamiltonian operators introduced in Section 5.4.3.

6.3 AUTOMATIC SYMMETRISATION

An efficient, numerically motivated method for constructing a symmetry-adapted set of wavefunctions is based on the so-called 'eigenfunction' method, which involves a diagonalisation of matrix representations of some operators $\hat{A}$, chosen to be invariant to all symmetry operations $R \in \boldsymbol{G}$. As we will show, a ro-vibrational Hamiltonian operator $\hat{H}$ can also be used for the 'eigenfunction' method. Indeed, $\hat{H}$ has the right property to commute with any symmetry operation R from $\boldsymbol{G}$

$$[\hat{H}, R] = 0 \tag{6.20}$$

and thus to share its eigenfunctions (up to a linear combination of degenerate states) and transform as one of the irreps of the system.

Here we aim to select and eigen-solve some suitable reduced Hamiltonian operators and use the eigenfunction technique for symmetrisation of a ro-vibrational basis set. More specifically, we will use reduced operators $\hat{H}^{(i)}$ introduced in Section 5.4.3 (see, e.g., Eq. (5.40)). These operators constructed by integrating $\hat{H}$ over all other degrees of freedom, satisfy Eq. (6.20) and therefore fit this purpose perfectly. Hence, by solving the reduced eigenvalue problem for each subspace i ($i = 1 \ldots L$) as given by

$$\hat{H}^{(i)}(\boldsymbol{Q}^{(i)})\Psi^{(i)}_{\lambda_i}(\boldsymbol{Q}^{(i)}) = E_{\lambda_i}\Psi^{(i)}_{\lambda_i}(\boldsymbol{Q}^{(i)}), \tag{6.21}$$

we obtain eigenfunctions $\Psi^{(i)}_{\lambda_i}$ that necessarily transform according to an irrep Γ_s of $\boldsymbol{G}$. Thus not only do we obtain a more compact basis set representation which can be efficiently contracted following the diagonalisation/truncation approach, it is also automatically symmetrised. In Eq. (6.21), E_{λ_i} is an eigenvalue associated with the eigenfunction $\Psi^{(i)}_{\lambda_i}$ and λ_i counts all the solutions from the subspace i and $\boldsymbol{Q}^{(i)}$ is a vector representing a set of vibrational coordinates $\{q_k, q_l, \ldots\}$ from a given subspace i. In our procedure, a subspace is built from equivalent vibrational coordinates, i.e. that transform to each other according to the symmetry operations of the group, e.g. r_1, r_2, r_3 or $\alpha_1, \alpha_2, \alpha_3$ in the case of PH_3 shown in Fig. 6.3.

6.4 NUMERICAL EXAMPLES

6.4.1 Symmetry-adapted vibrational basis set for an XY_2-type molecule

As a practical demonstration of the eigenfunction symmetrisation methodology, let us symmetrise the vibrational basis functions for a XY_2 triatomic molecule using

Table 6.5 Transformation properties of the internal coordinates r_1, r_2 and α of an XY_2-type molecule and the characters of the irreps of the $\mathcal{C}_{2v}(M)$ group.

	Operations			
Coordinate	E	(12)	E^*	$(12)^*$
r_1	r_1	r_2	r_1	r_2
r_2	r_2	r_1	r_2	r_1
α	α	α	α	α
Irrep Γ	Characters χ			
A_1	1	1	1	1
A_2	1	1	-1	-1
B_1	1	-1	-1	1
B_2	1	-1	1	-1

the primitive basis set functions $|v_1\rangle|v_2\rangle|v_3\rangle = \phi_{v_1}(r_1)\phi_{v_2}(r_2)\phi_{v_3}(\alpha)$ and thus obtain the well-known symmetry-adapted combinations of XY_2 vibrational basis functions as given by (for $n_1 \neq n_2$)

$$\begin{aligned}\Phi^{A_1}_{n_1,n_2,n_3} &= \frac{1}{\sqrt{2}}\left[\phi_{n_1}(r_1)\phi_{n_2}(r_2) + \phi_{n_2}(r_1)\phi_{n_1}(r_2)\right]\phi_{n_3}(\alpha), \\ \Phi^{B_2}_{n_1,n_2,n_3} &= \frac{1}{\sqrt{2}}\left[\phi_{n_1}(r_1)\phi_{n_2}(r_2) - \phi_{n_2}(r_1)\phi_{n_1}(r_2)\right]\phi_{n_3}(\alpha)\end{aligned} \tag{6.22}$$

and (for $n_1 = n_2 \equiv n$)

$$\Phi^{A_1}_{n,n,n_3} = \phi_n(r_1)\phi_n(r_2)\phi_{n_3}(\alpha), \quad n_1 = n_2 \equiv n.$$

Here A_1 and B_2 are two irreducible representations of $\mathcal{C}_{2v}(M)$ (see Table 6.5).

The vibrational modes of XY_2 span the Abelian group $\mathcal{C}_{2v}(M)$ with two equivalent stretching modes 1 and 2, whose transformation properties are associated with the permutation (12). The 'reducible' primitive functions $|v_1\rangle|v_2\rangle|v_3\rangle$ (for $n_1 \neq n_2$) transform via the permutation of v_1 and v_2 as follows:

$$(12)|v_1\rangle|v_2\rangle|v_3\rangle = |v_2\rangle|v_1\rangle|v_3\rangle, \tag{6.23}$$

$$(12)|v_2\rangle|v_1\rangle|v_3\rangle = |v_1\rangle|v_2\rangle|v_3\rangle. \tag{6.24}$$

Let us consider a numerical example of vibrational wavefunctions of the H_2S molecule obtained variationally as illustrated in Fig. 5.4. We first construct matrix representations of the reduced Hamiltonians described in Section 5.4.3 for two subspaces, stretching (modes 1,2) and bending (mode 3) as given by

$$\hat{H}^{(1)}(r_1, r_2) = \langle 0_3|\hat{H}|0_3\rangle, \tag{6.25}$$

$$\hat{H}^{(2)}(\alpha) = \langle 0_1|\langle 0_2|\hat{H}|0_2\rangle|0_1\rangle. \tag{6.26}$$

For simplicity here we use a small basis set limited by the polyad number $P_{\max} = 2$, i.e.

$$P = v_1 + v_2 + v_3 \leq 2.$$

After solving the reduced eigenvalue problem for $\hat{H}^{(1)}$ in Eq. (6.25) numerically, the following variational wavefunctions are obtained:

$$\Psi_1^{(1)}(r_1, r_2) = 0.99999|0,0\rangle + 0.000055\,(|0,1\rangle + |1,0\rangle) + \ldots, \quad (6.27)$$
$$\Psi_2^{(1)}(r_1, r_2) = 0.000078|0,0\rangle - 0.707107\,(|0,1\rangle + |1,0\rangle) + \ldots, \quad (6.28)$$
$$\Psi_3^{(1)}(r_1, r_2) = -0.707107\,(|0,1\rangle - |1,0\rangle) + \ldots, \quad (6.29)$$

where the shorthand notation $|v_1, v_2\rangle = |v_1\rangle|v_2\rangle$ is used. Let us note that $\Psi_1^{(1)}(r_1, r_2)$ has a dominant basis set contribution from $|0,0\rangle$ and can be assigned according to $\Psi_{0,0}(r_1, r_2)$, while $\Psi_2^{(1)}$ and $\Psi_3^{(1)}$ belong to the fundamentals, symmetric and asymmetric, respectively (see Section 5.4.5).

The numerical eigenfunctions $\Psi_1^{(1)}$ and $\Psi_2^{(1)}$ have the expected symmetrised form as in Eq. (6.22) and also transform according to the group characters (see Table 6.5) as given by

$$(12)\,\Psi_1^{(1)}(r_1, r_2) = \Psi_1^{(1)}(r_1, r_2), \quad (6.30)$$
$$(12)\,\Psi_2^{(1)}(r_1, r_2) = \Psi_2^{(1)}(r_1, r_2), \quad (6.31)$$
$$(12)\,\Psi_3^{(1)}(r_1, r_2) = -\Psi_3^{(1)}(r_1, r_2). \quad (6.32)$$

With these properties, we classify them as $\Psi_1^{(1),A_1}$, $\Psi_2^{(1),A_1}$ and $\Psi_3^{(1),B_2}$, respectively.

The 1D eigenfunctions of subset 2 (bending mode) have the trivial symmetry A_1 by construction. The 3D symmetry-adapted vibrational basis set is then formed as a product of wavefunctions $\Psi_{\lambda_1}^{(1),\Gamma_s}(r_1, r_2)$ and $\Psi_{\lambda_2}^{(2),\Gamma_s'}(\alpha)$ with the symmetry inherited from subset 1.

It should be noted that $\Psi_1^{(1)}$ and $\Psi_2^{(1)}$ were obtained numerically in the symmetry-adapted form without any assumption on their symmetries and their irreps were deduced by examining their numerical expansion coefficients $T_{k,\nu,n}^{\mu,J,\Gamma_s} = T_{\{n_1,n_2\}}^{\mu,\Gamma_s}$ as in Eq. (6.10). Here $J = 0, k = 0$ (rotational indices) and $n = 1$ (degenerate component) are omitted for simplicity and $\nu = \{n_1, n_2\}$. The numerical error of the symmetrisation was obtained within 10^{-15}, estimated by comparing $T_{\{1,0\}}^{1,A_1}$ to $T_{\{0,1\}}^{1,A_1}$, $T_{\{1,0\}}^{2,A_1}$ to $T_{\{0,1\}}^{2,A_1}$ and $T_{\{1,0\}}^{3,B_2}$ to $-T_{\{0,1\}}^{3,B_2}$. This analysis based on the manual inspection of $T_{k,\nu,n}^{\mu,J,\Gamma_s}$ was trivial in the case of XY_2. Our goal, however, is an *automatic* numerical symmetrisation algorithm applicable to arbitrary basis sets, coordinates, symmetries or molecules. As will be demonstrated below, the importance of an automatic symmetrisation procedure becomes more pronounced for larger molecules with more complicated symmetries, especially for ones containing degenerate representations.

6.4.2 Symmetry-adapted vibrational basis set for an XY_3 molecule, $\mathcal{C}_{3v}$-symmetry

A more complicated and interesting example is symmetry adaptation of the vibrational basis set for XY_3, a rigid pyramidal tetratomic molecule which spans the $\mathcal{C}_{3v}$(M) molecular symmetry group.

Using the product basis set for the primitive stretching functions of XY_3 as in Eq. (6.3):

$$\phi_{v_1,v_2,v_3}(r_1,r_2,r_3) = |v_1\rangle|v_2\rangle|v_3\rangle$$

we follow the eigenfunction method and solve the stretching problem in Eq. (5.39) (subset 1) for the corresponding reduced Hamiltonian as given in Eq. (5.40). The associated permutation properties and group characters of $\mathcal{C}_{3v}$(M) are listed in Tables 6.1 and 6.2. The $\mathcal{C}_{3v}$(M) group contains the $\Gamma_s = A_1$, A_2 and E representations, where E is two fold degenerate. According to the property of the symmetry operators to commute with $\hat{H}$ (see Eq. 6.20), the resulting wavefunctions $\Psi^{(i)}_{\lambda_i}$ are expected to be eigenfunctions of all six symmetry operators R of $\mathcal{C}_{3v}$(M) from Table 6.1 and therefore to necessarily transform as A_1, A_2 or E.

Using the PH_3 example from Section 5.3.5, we form a primitive (non-symmetrised) stretching basis set using the polyad number cut-off of $P_{\rm max} = 10$, i.e.

$$P = v_1 + v_2 + v_3 \leq P_{\rm max} = 10.$$

After solving the eigen-problem for the reduced stretching Hamiltonian $\hat{H}^{(1)}_{\rm str}$ numerically, the following variational eigenfunctions corresponding to the lowest four eigenvalues were obtained (where the shorthand notation $|v_1,v_2,v_3\rangle \equiv |v_1\rangle|v_2\rangle|v_3\rangle$ is used)

$$\begin{aligned}
\Psi^{(1)}_1 &= 0.9997|0,0,0\rangle - 0.128\,(|1,0,0\rangle + |0,1,0\rangle + |0,0,1\rangle) + \dots \\
\Psi^{(1)}_2 &= -0.0223|0,0,0\rangle - 0.57689\,(|1,0,0\rangle + |0,1,0\rangle + |0,0,1\rangle) + \dots \\
\Psi^{(1)}_{3,1} &= 0.50667|0,0,1\rangle - 0.80753|0,1,0\rangle + 0.30086|1,0,0\rangle + \dots \\
\Psi^{(1)}_{3,2} &= 0.63993|0,0,1\rangle + 0.11883|0,1,0\rangle - 0.75875|1,0,0\rangle + \dots .
\end{aligned} \tag{6.33}$$

The corresponding eigenvalues (energy term values) are 0.0, 2317.86, 2328.28 and 2328.28 cm^{-1}, respectively, relative to the zero-point energy (ZPE) of 5222.59 cm^{-1}.

Based on the largest basis set contribution approach, $\Psi^{(1)}_1$ is the vibrational ground state $(0,0,0)$, while $\Psi^{(1)}_2$, $\Psi^{(1)}_{3,1}$ and $\Psi^{(1)}_{3,2}$ represent fundamental states. $\Psi^{(1)}_1$ and $\Psi^{(1)}_2$ in Eq. (6.33) have fully symmetric form and thus belong to A_1. The irreps of $\Psi^{(1)}_{3,1}$ and $\Psi^{(1)}_{3,2}$ cannot be immediately recognised from their expansion coefficients, but the coinciding energy levels indicate a degenerate solution, which for the case of $\mathcal{C}_{3v}$(M) can only mean the E symmetry. We therefore introduce the subscript λ, n for these two wavefunctions to label the degenerate components ($n = 1, 2$) of the state $\lambda = 3$.

All four wavefunctions (as well as other solutions not shown here) appear readily symmetrised as expected. However, guessing the degenerate E symmetries based on the degeneracy of energies, which worked here, is not sufficiently robust and even reliable for practical numerical calculations, especially for high excitations with accidentally close energies (accidental resonances). The A_2 states as single energy solutions can be also easily be mixed up with A_1.

Finally, the degenerate components usually come out of diagonalisations as arbitrary degenerate mixtures and therefore do not necessarily obey standard irreducible transformation rules, which is a desirable property. For example, we expect the degenerate pair $(\Psi^{(1)}_{3,1}, \Psi^{(1)}_{3,2})$ in Eq. (6.33) to transform with $\mathbf{D}[(123)]$ according to Eq. (6.17), which is not the case. Therefore, these randomly mixed degenerate components have to be further unitarily transformed to the standard E-symmetry form.

To conclude this section, the eigenfunction symmetrisation method is a robust approach for symmetry basis set adaption, suitable for numerical implementations. However, this method does not automatically reveal which irreps these functions belong to. Besides, the degenerate components are randomly mixed, which is not ideal for subsequent applications. This is where the symmetry sampling approach comes in, as will be described in the next section.

6.5 SYMMETRY SAMPLING OF EIGENFUNCTIONS

Here we show how to reconstruct symmetries Γ_s of the eigenfunctions $\Psi^{(i)}_{\lambda_i,n}$ obtained using the eigenfunction approach as given in Eq. (6.21) and how to bring their degenerate components n into the 'standard' form, all in an automatic manner. To this end, we analyse the symmetry properties of the eigenfunctions $\Psi^{(i)}_{\lambda_i,n}(\mathbf{Q}^{(i)})$ by sampling them on a grid of instantaneous nuclear configurations.

Let us consider an l_λ-fold set of degenerate eigenfunctions $\Psi_{\lambda,n}$ $(n = 1, \ldots, l_\lambda)$ from a subspace i and define a grid of randomly selected geometries $\mathbf{Q}^{(i)}_k$ $(k = 1 \ldots N^{(i)}_{\rm grid})$. We assume that the transformation properties of the coordinates with respect to R are known at any specific point k. This can be expressed as

$$R\,\mathbf{Q}^{(i)}_k = \mathbf{Q}'^{(i)}_k. \tag{6.34}$$

Each subspace is independent from the others by definition, and we omit the superscript (i) unless required. Let us assume that we can evaluate $\Psi_{\lambda,n}(\mathbf{Q})$ at any grid point k to obtain

$$\begin{aligned}\Psi_{\lambda,n}(k) &\equiv \Psi_{\lambda,n}(\mathbf{Q}_k), \\ \Psi'_{\lambda,n}(k) &\equiv R\Psi_{\lambda,n}(\mathbf{Q}_k) = \Psi_{\lambda,n}(\mathbf{Q}'_k),\end{aligned} \tag{6.35}$$

where $\mathbf{Q}'$ and Ψ' are the transformed coordinates and functions, respectively. The eigenfunctions $\Psi_{\lambda,n}(\mathbf{Q}_k)$ and their symmetric images $R\,\Psi_{\lambda,n}(\mathbf{Q}_k)$ are related via the transformation matrices as given by

$$R\Psi_{\lambda,m}(\mathbf{Q}_k) = \sum_{n=1}^{l_\lambda} D_{m,n}[R]\Psi_{\lambda,n}(\mathbf{Q}_k), \tag{6.36}$$

where the passive transformation convention was used; see Section 6.7. For example, for the E-symmetry wavefunctions from Eq. (6.33), we obtain

$$\begin{pmatrix} \Psi'_{3,1}(k) \\ \Psi'_{3,2}(k) \end{pmatrix} = R \begin{pmatrix} \Psi_{3,1}(k) \\ \Psi_{3,2}(k) \end{pmatrix} = \begin{pmatrix} D_{11}[R] & D_{12}[R] \\ D_{21}[R] & D_{22}[R] \end{pmatrix} \times \begin{pmatrix} \Psi_{3,1}(k) \\ \Psi_{3,2}(k) \end{pmatrix}. \tag{6.37}$$

By combining Eq. (6.35) and Eq. (6.36), we obtain

$$\sum_{n=1}^{l_\lambda} D_{mn}[R]\Psi_{\lambda,n}(k) = \Psi'_{\lambda,m}(k). \tag{6.38}$$

Here the degeneracy parameter l_λ is not the degeneracy order of irreducible representations Γ_s (2 for E or 3 for F), but the number of degenerate reducible states. This value can be deduced by simply taking the number of states with the same energies (within some threshold). For non-degenerate wavefunctions ($l_\lambda = 1$), the sampling procedure will always produce $D_{1,1}[R] = \chi[R] = \pm 1$. Accidental degeneracies (e.g. A_1 or A_2 states of identical energies) can be processed as if they were normal degenerate components which should lead to diagonal representation matrices $D_{i,i}[R] = \pm 1$ and $D_{i,j}[R] = 0$ $(i \neq j)$.

If $\Psi_{\lambda,n}(\mathbf{Q}_k)$ and their images $\Psi'_{\lambda,m}(k)$ are known, $(l_\lambda)^2$ elements $D_{mn}[R]$ can be determined by solving Eq. (6.38) as a system of N_{grid} linear equations ($k = 1 \ldots N_{\text{grid}}$) of the type

$$\sum_n A_{kn} x_n^{(m)} = b_k^{(m)}, \tag{6.39}$$

where $A_{kn} = \Psi_{\lambda,n}(k)$, $b_k^{(m)} = \Psi'_{\lambda,m}(k)$ and $x_n^{(m)} = D_{nm}[R]$. It should be noted that these equations need to be supplemented by the ortho-normality condition in order for matrices $\boldsymbol{D}[R]$ to be unitary. In principle it is also sufficient to include in Eq. (6.38) only operations R representing the group generators, which can significantly reduce its size. Once all the g transformation matrices $\mathbf{D}[R]$ have been found (g is the group order), the standard projection operator approach is applied to generate the irreducible representations (see Section 6.2).

At least $N_{\text{grid}} = (l_\lambda)^2$ grid points are required to define the linear system given in Eq. (6.39) (or even fewer due to the unitary property of the transformation matrices). In practice, it is difficult to find a proper set of geometries with all values of $\Psi'_\lambda(k)$ and $\Psi_\lambda(k)$ large enough to make the solution of this set of linear equations numerically stable (i.e. with a non-vanishing determinant). As a remedy, more points ($N_{\text{grid}} \gg (l_\lambda)^2$) can be selected. The resulting over-determined linear system can be solved using standard linear algebra libraries, e.g. using the singular value decomposition method.

The sampling symmetrisation procedure can be applied to any primitive functions provided their values can be calculated at any instantaneous geometry. It should be noted that it is different from the more common approach of directly exploring the permutational properties of the wavefunctions. The advantage of the sampling symmetrisation procedure is that by applying the group transformations to the coordinates instead of the basis functions it provides a more general numerical approach applicable to basis functions with no obvious permutation symmetries of the wavefunctions.

6.5.1 Sampling symmetrisation of a XY_3 vibrational basis set, the $\mathcal{C}_{3v}$-symmetry

Let us now return to the XY_3 example above and apply the sampling symmetrisation approach to the degenerate state $\Psi_{3,n}^{(1)}$ in Eq. (6.33). By solving Eq. (6.38) for all six $\mathcal{C}_{3v}$(M) group operations (see Table 6.1), the following transformation matrices were determined (using 40 sampling points):

$$\mathbf{D}[E] = \begin{pmatrix} 1.000 & 0.000 \\ 0.000 & 1.000 \end{pmatrix}, \quad \mathbf{D}[(123)] = \begin{pmatrix} -0.500 & 0.866 \\ -0.866 & -0.500 \end{pmatrix},$$
$$\mathbf{D}[(321)] = \begin{pmatrix} -0.500 & -0.866 \\ 0.866 & -0.500 \end{pmatrix}, \quad \mathbf{D}[(23)^*] = \begin{pmatrix} -0.581 & -0.814 \\ -0.814 & 0.581 \end{pmatrix},$$
$$\mathbf{D}[(13)^*] = \begin{pmatrix} 0.995 & -0.097 \\ -0.097 & -0.995 \end{pmatrix}, \quad \mathbf{D}[(12)^*] = \begin{pmatrix} -0.414 & 0.910 \\ 0.910 & 0.414 \end{pmatrix}.$$

In principle only matrices corresponding to the group generators should suffice (three in this case), but here all operations are used. The group characters $\chi^{\Gamma}[R]$ of these transformations are 2.0, −1.0 and 0.0 (within given numerical precision). Using Eq. (6.12), we obtain the following reduction coefficients $a^E = 1$ and $a^{A_1} = a^{A_2} = 0$ $(\pm 10^{-12})$ with a single E- symmetry component as expected for a doubly degenerate solution.

We are now ready to symmetrise the wavefunction $\Psi_{3,n}^{(1)}$ from Eq. (6.33). We first build a projection operator P_{11}^{E} given in Eq. (6.16) using $\mathbf{D}[R]$ from above, which is then applied to the degenerate components $\Psi_a = \Psi_{3,1}^{(1)}$ and $\Psi_b = \Psi_{3,2}^{(1)}$ to obtain

$$\tilde{\Psi}_a = 0.1359\,\Psi_a - 0.3426\,\Psi_b,$$

which after normalisation becomes

$$\tilde{\Psi}_a = 0.3686\,\Psi_a - 0.9296\,\Psi_b.$$

The second component $\tilde{\Psi}_b$ is found by applying the transfer operator from Eq. (6.16):

$$\tilde{\Psi}_b = \frac{2}{6}\sum_R D^{\Gamma_s}[R]_{12}^{*}\tilde{\Psi}_a.$$

After normalisation, we obtain $\tilde{\Psi}_b = 0.9296\,\Psi_a + 0.3686\,\Psi_b$. As mentioned above, if the projection operator $P_{11}^{\Gamma_s}$ does not lead to a correct or independent combination, a different component of $P_{mm}^{\Gamma_s}$ should be used until the correct solution is found (which is guaranteed).

By taking into account the original Ψ_a and Ψ_b from Eq. (6.33), the final symmetry-adapted representation for $\tilde{\Psi}_3^{(1)}$ is obtained:

$$\tilde{\Psi}_{3,1}^{(1)} = \tilde{\Psi}_{E_a}^{(1)} = -\frac{1}{\sqrt{6}}\left(\,|0\rangle|0\rangle|1\rangle + |0\rangle|1\rangle|0\rangle - 2\,|1\rangle|0\rangle|0\rangle\,\right) + \ldots \quad (6.40)$$

$$\tilde{\Psi}_{3,2}^{(1)} = \tilde{\Psi}_{E_b}^{(1)} = \frac{1}{\sqrt{2}}\left(\,|0\rangle|0\rangle|1\rangle - |0\rangle|1\rangle|0\rangle\,\right) + \ldots \quad (6.41)$$

This is a well-known expression for the standard E-symmetry representations of $\mathcal{C}_{3\mathrm{v}}(\mathrm{M})$ wavefunctions reproducing symmetry-adapted expressions given in Eqs. (6.6, 6.7). The expansion coefficients in Eqs. (6.40, 6.41) are typically defined within a numerical error of $\sim 10^{-14}$.

6.5.2 Reduction of a product

For our example of a rigid pyramidal tetratomic molecule XY_3, we can now apply the same automatic symmetry adaptation procedure to all eigenfunctions obtained by solving the reduced Schrödinger equations corresponding to two independent vibrational subspaces (subspace 1 $\{\Delta r_1, \Delta r_2, \Delta r_3\}$ and subspace 2 $\{\Delta\alpha_{12}, \Delta\alpha_{13}, \Delta\alpha_{23}\}$) as given in Eq. (5.40). As a result, two subsets of contracted wavefunctions $\Psi^{(1)}_{\lambda_1}$ and $\Psi^{(2)}_{\lambda_2}$ are obtained, each of which is transformed according to A_1, A_2 or E.

Once all symmetry-adapted eigenfunctions for each subspace $i = 1, 2$ have been found, the final vibrational basis set is formed as a direct product

$$\Psi^{\Gamma_1,\Gamma_2}_{\lambda_1,\lambda_2} = \Psi^{(1),\Gamma_1}_{\lambda_1} \otimes \Psi^{(2),\Gamma_2}_{\lambda_2},$$

which is not irreducible and has to be further symmetrised using, e.g., the projection/transfer operator approach described above; see Eqs. (6.14, 6.16). The required transformation matrices are obtained as products of standard irreducible transformation matrices, commonly tabulated for point groups.

$$\boldsymbol{D}^{\Gamma_1,\Gamma_2}[R] = \boldsymbol{D}^{\Gamma_1}[R]\,\boldsymbol{D}^{\Gamma_2}[R].$$

For instance, consider a product of two degenerate functions $\Psi^{(1),E}_{n_1} \otimes \Psi^{(2),E}_{n_2}$, which corresponds to four combinations $\Psi^{(1),E_\alpha}_{n_1} \times \Psi^{(2),E_\beta}_{n_2}$ $(\alpha, \beta = a, b)$:

$$\begin{aligned} &\Psi^{E_a}_{n_1} \times \Psi^{E_a}_{n_2}, \qquad && \Psi^{E_a}_{n_1} \times \Psi^{E_b}_{n_2}, \\ &\Psi^{E_b}_{n_1} \times \Psi^{E_a}_{n_2}, && \Psi^{E_b}_{n_1} \times \Psi^{E_b}_{n_2}. \end{aligned}$$

Here n_1 and n_2 are vibrational indices (quantum numbers) of the subspace 1 and 2, respectively, and a and b are two degenerate components (some indexes are omitted for clarity). These four components transform as a direct product of E-representation matrices

$$\boldsymbol{D}^{\Gamma_s}[R] = \boldsymbol{D}^{E}[R] \otimes \boldsymbol{D}^{E}[R],$$

which forms a reducible representation with the characters obtained as a product of the corresponding characters of $\boldsymbol{D}^{E}[R]$:

$$\chi^{\Gamma_s}[R] = \chi^{E} \times \chi^{E}.$$

For $\mathcal{C}_{3\mathrm{v}}(\mathrm{M})$, we obtain $\chi^{\Gamma_s} = 4$, 1 and 0 for E, (123) and (12)*, respectively; see Table 6.1. In fact, this is the standard textbook example of a reduction of the $E \otimes E$

product. The reduction coefficients are 1, 1 and 1 for A_1, A_2 and E as obtained from Eq. (6.12), i.e. the product of two E irreps is reduced to three irreps

$$E \otimes E = A_1 \oplus A_2 \oplus E.$$

Using the projection technique, the reducible products of irreducible functions are reduced to

$$\Psi^{A_1}_{n_1,n_2} = \frac{1}{\sqrt{2}}\left[\Psi^{(E_a,E_a)}_{n_1,n_2} + \Psi^{(E_b,E_b)}_{n_1,n_2}\right], \tag{6.42}$$

$$\Psi^{A_2}_{n_1,n_2} = \frac{1}{\sqrt{2}}\left[\Psi^{(E_a,E_b)}_{n_1,n_2} - \Psi^{(E_b,E_a)}_{n_1,n_2}\right], \tag{6.43}$$

$$\Psi^{E_a}_{n_1,n_2} = \frac{1}{\sqrt{2}}\left[\Psi^{(E_a,E_a)}_{n_1,n_2} + \Psi^{(E_b,E_b)}_{n_1,n_2}\right], \tag{6.44}$$

$$\Psi^{E_b}_{n_1,n_2} = \frac{1}{\sqrt{2}}\left[\Psi^{(E_a,E_b)}_{n_1,n_2} - \Psi^{(E_b,E_a)}_{n_1,n_2}\right], \tag{6.45}$$

where the following shorthand notation is used:

$$\Psi^{(\Gamma,\Gamma')}_{n_1,n_2} = \Psi^{(1),E}_{n_1,l} \times \Psi^{(2),E}_{n_2,l'}.$$

6.5.3 Symmetry properties of rigid rotor wavefunctions

The symmetrisation of the rigid rotor wavefunctions $|J,k,m\rangle$ is usually straightforward. It can be shown that for most of the systems (belonging to, e.g., $\mathcal{C}_{n\mathrm{v}}$, $\mathcal{C}_{n\mathrm{h}}$, $\mathcal{D}_{n\mathrm{h}}$ or $\mathcal{D}_{n\mathrm{d}}$) the Wang-type combinations of $|k|$ and $-|k|$ in Eq. (5.82) already have correct forms that transform according to some irreps of the given symmetry group. Moreover, their symmetries are directly characterised by the corresponding values $|k|$. This is possible because all symmetry operations from these groups can be associated with some equivalent rotations about the body-fixed axes x, y and z only. In this case, the group operations R can be represented by the two types of equivalent rotations, R^{π}_{α} and R^{β}_{z}: R^{π}_{α} is a rotation of the molecule-fixed xyz axes through π radians about an axis in the xy plane making an angle α with the x axis (α is measured in the right-hand rule about the z axis), and R^{β}_{z} is a rotation of the molecule-fixed axes through β radians about the z axis (measured in the right-handed sense about the z axis). As an illustration, equivalent rotations for $\mathcal{C}_{2\mathrm{v}}$(M) and $\mathcal{C}_{3\mathrm{v}}$(M) are listed in Table 6.7.

Consider, for example, a triatomic molecule XY_2 shown in Fig. 6.4 with the x molecular-fixed axis bisecting the bond angle Y–X–Y, the z axis pointing from atom 1 to atom 2 and the y axis chosen in the right-handed sense. The equivalent rotations for the operations E, (12), E^* and (12)* ($\mathcal{C}_{2\mathrm{v}}$(M)) are R^0, R^{π}_{x}, R^{π}_{y} and R^{π}_{z}, respectively; see Table 6.7. Here R^{π}_{x} and R^{π}_{y} are rotations of the molecule-fixed (x, y and z) axes through π radians about an axis in the xy plane making angles 0 and $\pi/2$ with the x axis.

For a rigid XY_3 molecule with the $\mathcal{C}_{3\mathrm{v}}$(M) group, there are two types of equivalent rotations: two rotations about the z axis through $2\pi/3$ and $4\pi/3$ radians and

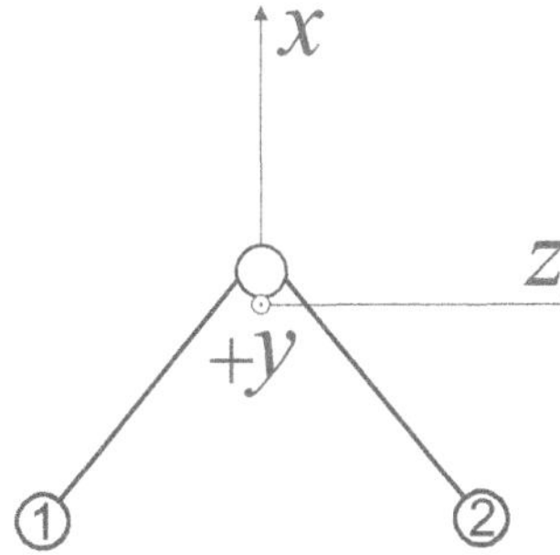

Figure 6.4 An equilibrium xyz frame of XY_2. All three axes x, y and z define equivalent rotations of $\mathcal{C}_{2v}(M)$.

three π radians rotations about the lines orthogonal to z (i.e. in the xy plane) and lying in the planes z-X-Y_i (see Fig. 6.5), where the equilibrium structure is assumed.

The rigid rotor wavefunctions can be shown to have the following properties upon these two types of equivalent rotations:

$$R_z^{\beta}|J,k,m\rangle = e^{ik\beta}|J,k,m\rangle, \tag{6.46}$$
$$R_{\alpha}^{\pi}|J,k,m\rangle = (-1)^J e^{-2ik\alpha}|J,-k,m\rangle. \tag{6.47}$$

Similar transformation properties can be derived for the parity-defined Wang functions in Eqs. (5.83, 5.84), which we reproduce here for convenience:

$$\begin{aligned}
|J,0,\tau\rangle &= |J,0\rangle, \quad \tau = J \bmod 2, \\
|J,K,\tau=0\rangle &= \frac{1}{\sqrt{2}}\left[|J,K,m\rangle + (-1)^{J+K}|J,-K,m\rangle\right], \\
|J,K,\tau=1\rangle &= \frac{i(-1)^{\sigma}}{\sqrt{2}}\left[|J,K,m\rangle - (-1)^{J+K}|J,-K,m\rangle\right].
\end{aligned} \tag{6.48}$$

Here $K = |k|$, τ is the value associated with the parity of $|J,K,\tau\rangle$ and m is omitted on the left-hand side for simplicity's sake. The meaning of σ is explained below. K is commonly used as the rotational quantum number, K_a or K_c, depending on the orientation of the z axis. The sign of k is, however, not a physically meaningful quantity of a rotational eigen-state, but the parity τ is.

Let us consider our example of XY_2 and reconstruct symmetries of the symmetrised rotational Wang functions in Eq. (6.48). By applying the four symmetry operations from Table 6.5 to $|J,K,\tau\rangle$ via their equivalent rotations in Eqs. (6.46–6.47), we obtain

$$E|J,k,m\rangle = |J,k,m\rangle, \tag{6.49}$$
$$(12)|J,k,m\rangle = R_0^{\pi}|J,K,\tau\rangle = (-1)^J|J,-k,m\rangle, \tag{6.50}$$
$$E^*|J,k,m\rangle = R_{\pi/2}^{\pi}|J,K,\tau\rangle = (-1)^{J+K}|J,-k,m\rangle, \tag{6.51}$$
$$(12)^*|J,k,m\rangle = R_z^{\pi}|J,K,\tau\rangle = (-1)^K|J,k,m\rangle. \tag{6.52}$$

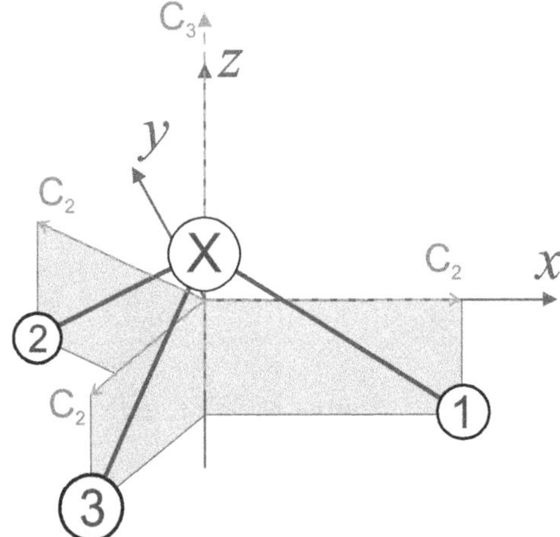

Figure 6.5 An equilibrium xyz frame of an XY_3 molecule and rotational axes C_2 and C_3 defining equivalent rotations of $\mathcal{C}_{3v}$(M).

The corresponding Wang wavefunctions transform as follows ($K \neq 0$):

$$E|J,K,\tau\rangle = |J,K,\tau\rangle \tag{6.53}$$

$$(12)|J,K,\tau\rangle = (-1)^J e^{-2iK\,0}|J,K,\tau\rangle = (-1)^{K+\tau}|J,K,\tau\rangle \tag{6.54}$$

$$E^*|J,K,\tau\rangle = e^{iK\pi/2}|J,K,\tau\rangle = (-1)^{\tau}|J,K,\tau\rangle \tag{6.55}$$

$$(12)^*|J,K,\tau\rangle = e^{iK\pi}|J,K,\tau\rangle = (-1)^{K}|J,K,\tau\rangle. \tag{6.56}$$

Comparing these properties to the group characters in Table 6.5, the $\mathcal{C}_{2v}$ irreps of the XY_2 Wang functions in Table 6.6 are obtained, for each pair of K and τ.

Table 6.6 Irreducible representations of $\mathcal{C}_{2v}$ of Wang functions in Eq. (5.82) for XY_2 for different pairs of K and τ.

K	τ	Γ_{rot}	K	τ	Γ_{rot}
even	0	A_1	odd	0	B_2
even	1	B_1	odd	1	A_2

For a general case, the arbitrary equivalent rotations R_α^π and R_z^β applied to $|J,K,\tau\rangle$ ($K \neq 0$, $\tau = 0, 1$) in Eq. (6.48) can be expressed as follows:

$$R|J,K,\tau\rangle = \sum_{\tau'=0}^{1} D_{\tau,\tau'}[R]|J,K,\tau\rangle, \tag{6.57}$$

where

$$\boldsymbol{D}[R_z^\beta] = \begin{pmatrix} \cos(K\beta) & (-1)^\sigma \sin(K\beta) \\ -(-1)^\sigma \sin(K\beta) & \cos(K\beta) \end{pmatrix}, \tag{6.58}$$

$$\boldsymbol{D}[R_\alpha^\pi] = (-1)^K \begin{pmatrix} \cos(2K\alpha) & (-1)^\sigma \sin(2K\alpha) \\ -(-1)^\sigma \sin(2K\alpha) & -\cos(2K\alpha) \end{pmatrix} \tag{6.59}$$

and $\boldsymbol{D}[R]$ is the corresponding transformation matrix in the representation of $|J, K, \tau\rangle$.

Table 6.7 Equivalent rotations of the $\mathcal{C}_{2\mathrm{v}}$(M) and $\mathcal{C}_{3\mathrm{v}}$(M) groups.

Symmetry	Operations/Equiv. rotations					
$\mathcal{C}_{2\mathrm{v}}$(M)	E	(12)	E^*	$(12)^*$		
	R^0	R_x^π	R_y^π	R_z^π		
$\mathcal{C}_{3\mathrm{v}}$(M)	E	(123)	(321)	$(23)^*$	$(13)^*$	$(12)^*$
	R^0	$R_z^{2\pi/3}$	$R_z^{4\pi/3}$	$R_{\pi/2}^\pi$	$R_{\pi/3}^\pi$	$R_{5\pi/3}^\pi$

The meaning of the factor $(-1)^\sigma$ can be illustrated for the XY_3 case of the $\mathcal{C}_{3\mathrm{v}}$(M) symmetry group. Based on the characters of the transformations of the Wang functions, the corresponding symmetries depend on the following cases:

$$\begin{array}{lcl} |J, K = 3t, \tau = 0\rangle & \rightarrow & A_1 \\ |J, K = 3t, \tau = 1\rangle & \rightarrow & A_2 \\ |J, K \neq 3t, \tau = 0\rangle & \rightarrow & E_a \\ |J, K \neq 3t, \tau = 1\rangle & \rightarrow & E_b \end{array} \tag{6.60}$$

where t is a positive integer number.

Considering the transformation properties of the operation (123) with its equivalent rotation $R_z^{2\pi/3}$ (see Table 6.7), we obtain

$$\boldsymbol{D}[R_z^{2\pi/3}] = \begin{cases} \begin{pmatrix} 1 & 0 \\ 0 & 1 \end{pmatrix}, & K = 3t \\ \begin{pmatrix} -\frac{1}{2} & (-1)^\sigma \frac{\sqrt{3}}{2} \\ -(-1)^\sigma \frac{\sqrt{3}}{2} & -\frac{1}{2} \end{pmatrix}, & K = 3t - 1 \\ \begin{pmatrix} -\frac{1}{2} & -(-1)^\sigma \frac{\sqrt{3}}{2} \\ (-1)^\sigma \frac{\sqrt{3}}{2} & -\frac{1}{2} \end{pmatrix}, & K = 3t + 1. \end{cases} \tag{6.61}$$

While the $K = 3t$ case corresponds to the A_1 or A_2 cases (depending on τ), the two other cases belong to the E symmetry of $\mathcal{C}_{3\mathrm{v}}$(M), which should match the corresponding (123) matrix from Table 6.4:

$$\boldsymbol{D}^E[(123)] = \begin{pmatrix} -\frac{1}{2} & \frac{\sqrt{3}}{2} \\ -\frac{\sqrt{3}}{2} & \frac{1}{2} \end{pmatrix}.$$

For $K \neq 3t$, this can only be satisfied if we choose $\sigma = K \bmod 3$. Thus, the factor $(-1)^\sigma$ in Eq. (6.48) is introduced to match the transformation properties of the symmetrised rigid rotor wavefunctions $|J, K, \tau\rangle$ to that of the tabulated group symmetry for any values of K.

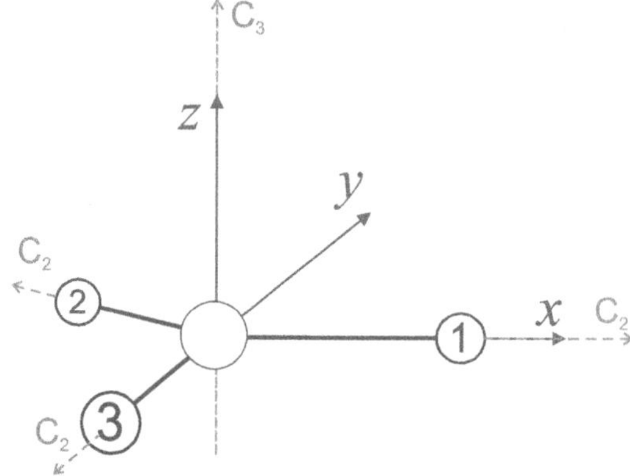

Figure 6.6 An equilibrium xyz frame of a planar XY_3 molecule and rotational axes C_2 and C_3 defining equivalent rotations of $\mathcal{D}_{3h}(M)$.

In order to show that $|J, K = 3t, \tau\rangle$ indeed transforms as A_1 or A_2 ($\tau = 0, 1$), let us apply the rotation $\boldsymbol{D}[R^{\pi}_{\pi/2}]$ (operation equivalent to $(23)^*$), using Eq. (6.59). For the case of $K = 3t$, we obtain

$$\boldsymbol{D}[R^{\pi}_{\pi/2}] = (-1)^K \begin{pmatrix} \cos(6\,t\,\pi) & (-1)^{\sigma} \sin(6\,t\,\pi) \\ -(-1)^{\sigma} \sin(6\,t\,\pi) & -\cos(6\,t\,\pi) \end{pmatrix} = \begin{pmatrix} 1 & 0 \\ 0 & -1 \end{pmatrix}.$$

Here we see that $|J, K = 3t, \tau = 0\rangle$ (1st row) is unchanged upon $(23)^*$ as A_1, while $|J, K = 3t, \tau = 1\rangle$ (2nd row) changes its sign as A_2.

As additional example, in Table 6.8 symmetry classifications of the rigid rotor wavefunction $|J, K, \tau\rangle$ $\mathcal{D}_{3h}(M)$ molecule XY_3 (e.g. non-rigid NH_3 or planar CH_3; see Fig. 6.6) and a $\mathcal{D}_{2h}(M)$ molecule X_2Y_4 (e.g. C_2H_4; see Fig. 6.7) are provided.

Table 6.8 Irreducible representations of the symmetrised wavefunction $|J, K, \tau\rangle$ in Eq. (6.48) for the $\mathcal{D}_{3h}(M)$ group of XY_3 (see Fig. 6.6) and the $\mathcal{D}_{2h}(M)$ group of X_2Y_4 (see Fig. 6.7) for different pairs of K and τ. Here t is a positive integer number, $\dagger = \prime$ if K is even and $\dagger = \prime\prime$ if K is odd.

XY_3, $\mathcal{D}_{3h}(M)$			X_2Y_4, $\mathcal{D}_{2h}(M)$		
K	τ	Γ_{rot}	K	τ	Γ_{rot}
$3t$	0	$A_1^{\dagger}$	even	0	A_g
$3t$	1	$B_1^{\dagger}$	even	1	B_{1g}
$3t \pm 1$	0	$E_a^{\dagger}$	odd	0	B_{3g}
$3t \pm 1$	1	$E_b^{\dagger}$	odd	1	B_{2g}

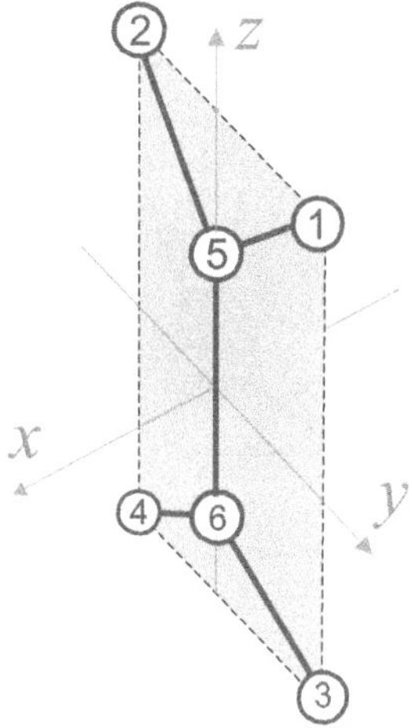

Figure 6.7 An xyz frame of the X_2Y_4 molecule.

6.6 SYMMETRISATION OF THE RIGID ROTOR BASIS FUNCTIONS: SPECIAL CASE

For some molecules, it is not possible to represent symmetrised rotational basis functions by simple Wang combinations of $|J,|k|,m\rangle$ and $|J,-|k|,m\rangle$. This is when the corresponding molecular group contains operations that cannot be described by equivalent rotations of the types R_{α}^{π} and R_{z}^{β} only.

Consider, e.g., a rigid XY_4 molecule spanning the $\mathcal{T}_d$(M) symmetry group (e.g. methane). One of the $\mathcal{T}_d$(M) group operations is the permutation (134) associated with the equivalent rotation $R_3(1,1,1)$ (see Fig. 6.8), which is a $2\pi/3$ right-hand rotation about the vector $(1,1,1)$. It can be shown that for this type of operations, two-component Wang-type functions in Eq. (6.48) do not transform irreducibly and therefore cannot serve as symmetry-adapted combinations. Instead, a correct symmetry adaptation requires a linear combination of $|J,k,m\rangle$ spanning a range of k values, which can be obtained using the standard projection protocol. Indeed, a reducible representation $\Gamma \in G$ can be reconstructed by applying the group operations on $|J,k,m\rangle$, which can be subsequently reduced to their irreducible combinations.

Symmetry transformation properties (i.e. transformation matrices $\boldsymbol{D}[R]$) of the rigid rotor wavefunctions required for the symmetrisation can be obtained directly from the Wigner D-functions, associated with the corresponding Euler angles of the particular equivalent rotation $R(\alpha,\beta,\gamma)$, which in the passive point of view are given by

$$R(\alpha,\beta,\gamma)|J,k,m\rangle = \sum_{k'=-J}^{J} D_{k',k}^{(J)*}(\alpha,\beta,\gamma)|J,k',m\rangle. \tag{6.62}$$

For example, for the operation $R_3(1,1,1)$ in XY_4, we obtain the Euler angles $(\pi,\pi/2,\pi/2)$ as an equivalent operation $R(\alpha,\beta,\gamma)$.

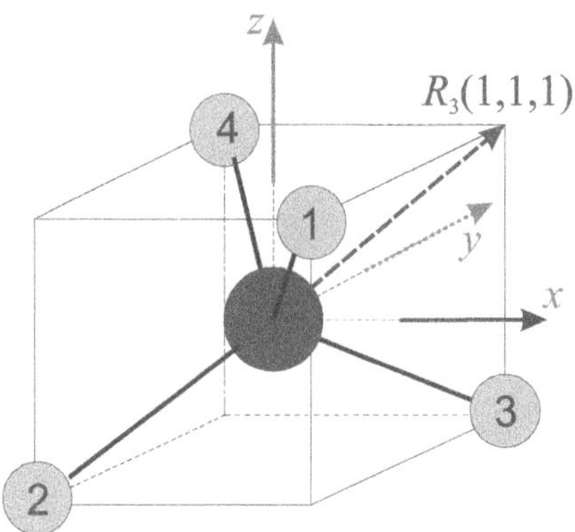

Figure 6.8 The equivalent rotation $R_3(1,1,1)$ for the (134) operation of $\mathcal{T}_{\rm d}(\mathrm{M})$ of the tetrahedral XY_4 molecule: a $\frac{2\pi}{3}$ right hand rotation about the vector from the origin to the point $(x,y,z)=(1,1,1)$.

The transformation properties of the multi-component Wang functions $|J,K,\tau\rangle$ in Eq. (6.48) can then be deduced via the unitary transformation $\boldsymbol{T}$

$$\boldsymbol{D}^{\rm Wang}[R] = \boldsymbol{T}^{+}\,\boldsymbol{D}^{(J)^*}(\alpha,\beta,\gamma)\,\boldsymbol{T},$$

where the $(2J+1)\times(2J+1)$ matrix $T_{i,j}$ is given by

$$T_{1,1} = \begin{cases} 1, & \text{even } J, \\ -i^J, & \text{odd } J \end{cases} \tag{6.63}$$

and

$$T_{n,n} = \frac{1}{\sqrt{2}}, \tag{6.64}$$

$$T_{n,n+1} = -i\frac{(-1)^{\sigma}}{\sqrt{2}}, \tag{6.65}$$

$$T_{n+1,n} = \frac{(-1)^{J+K}}{\sqrt{2}}, \tag{6.66}$$

$$T_{n+1,n+1} = i\frac{(-1)^{J+K+\sigma}}{\sqrt{2}}, \tag{6.67}$$

$$T_{n,n'} = 0, \quad \text{for} \quad |n-n'|>1, \tag{6.68}$$

where $n=2K$ and $K=1\ldots J$. Once the transformation matrices are known, the standard projection technique (Section 6.2) is applied to obtain the symmetry-adapted rigid rotor combinations used as rotational basis functions.

Since the symmetry-adapted rotational functions are obtained as linear combinations of $|J,k,m\rangle$ with $k=-J\ldots J$, the rotational quantum number K can no longer be used for classification of these symmetrised rigid rotor combinations. Instead they can be labelled as $|J,\Gamma,n\rangle$, where Γ is the symmetry and n is a counting

Table 6.9 Equivalent rotations of X_2Y_4 in terms of the Euler angles (α, β, γ) (Eq. (6.62)) of the molecular symmetry group $\mathcal{D}_{2h}$(M). The x, y, z equilibrium frame is as in Fig. 6.7.

R	Equiv. rot.	(α, β, γ)
E	R^0	$(0,0,0)$
$(12)(34)$	R_z^π	$(\pi,0,0)$
$(13)(24)(56)$	R_0^π	$(0,\pi,0)$
$(14)(23)(56)$	$R_{\pi/2}^\pi$	$(\pi/2,\pi,3\pi/2)$
E^*	$R_{\pi/2}^\pi$	$(\pi/2,\pi,3\pi/2)$
$(12)(34)^*$	R_0^π	$(0,\pi,0)$
$(13)(24)(56)^*$	R_z^π	$(\pi,0,0)$
$(14)(23)(56)^*$	R^0	$(0,0,0)$

index:

$$|J,\Gamma,n\rangle = \sum_k T_{n,k}^{(J,\Gamma)}|J,k,m\rangle. \tag{6.69}$$

It should be noted that this symmetry adaptation can also be easily formulated as a numerical procedure.

As a more general symmetrisation, this D-function-based technique (Eq. (6.62)) can also be applied to the Wang-type cases (Eq. (6.48)) presented above. For example, consider a rigid X_2Y_4 molecule of $\mathcal{D}_{2h}$(M). The equivalent rotations and corresponding Euler angles of X_2Y_4 required to obtain rotational representations $\boldsymbol{D}^{(J)^*}(\alpha, \beta, \gamma)$ are collected in Table 6.9. One of the symmetry operations, (12)(34) with the equivalent rotation R_a^π, corresponds to the rotation $R(\pi, 0, 0)$. Using Eq. (6.62), we obtain

$$R(\pi,0,0)^{(12)(34)}|J,k,m\rangle = e^{ik\pi}|J,k,m\rangle,$$

which is consistent with the definition given in Eq. (6.46).

6.7 MATERIALS USED

For the proof that the Hamiltonian matrix elements vanish between different symmetries, see Bunker and Jensen (1998).

Pauli principle is from Pauli (1925, 1940).

Origins of the molecular symmetry are by Longuet-Higgins (1963) and Hougen (1962b, 1963) and described by Bunker and Jensen (1998).

The automatic symmetrisation methodology and the sampling approach are from Yurchenko et al. (2017). For the eigenfunction method, see Jordanov and Orville-Thomas (1981), Chen et al. (1985) and Lemus (2003).

$\mathcal{T}_d$ symmetries of equivalent rotations of CH_4 are from Bunker and Jensen (1999). $\mathcal{T}_d$ symmetrisation of the rotational basis using CSCO is from Alvarez-Bajo et al. (2011). A full list of $\mathcal{T}_d$ equivalent rotations of CH_4 can be also found in Alvarez-Bajo et al. (2011).

The H_2S example is based on the PES by Azzam et al. (2016) and reproduced from Yurchenko et al. (2017).

The PH_3 example is based on the spectroscopic model from Sousa-Silva et al. (2015) and reproduced from Yurchenko et al. (2017).

When deriving transformation properties of degenerate wavefunctions, the convention by Bunker and Jensen (1998) to define the operations R on the nuclear coordinates and functions was used, which is also referred to as passive (see, e.g., a detailed discussion by Alvarez-Bajo et al. (2011)).

The reduction example used for building symmetry-adapted combinations of $\phi_1 = |v_1 v_1 v_3\rangle$, $\phi_2 = |v_1 v_3 v_1\rangle$ and $\phi_3 = |v_3 v_1 v_1\rangle$, in Section 6.2, closely follows Problem 6-2 from Bunker and Jensen (1998).

Point group irreducible transformation matrices can be found in Pfennig (2015).

MATERIALS NOT COVERED BUT RELEVANT

The symmetry-based method for quantum-number reconstruction is by Lemus (2012).

The symmetry operator used in the eigenfunction method can be constructed as a linear combination of symmetry operations from the so-called complete set of commuting operators (CSCO); see Chen et al. (1985) and Lemus (2003).

The example of symmetrisation using tensorial formalism (Zhilinskii et al., 1987; Champion et al., 1992; Boudon et al., 2004) with the standard numerically motivated symmetrisation approach using the projection technique is in Nikitin et al. (1997).

The symmetry-adapted method and KEO calculations are from Nikitin et al. (2015).

The symmetry adaptation in the Lanczos method in the calculation of molecular energies is from Wang and Carrington (2001).

For the extended molecule symmetry (EMS) groups, see Hougen (1964), Al-Refaie et al. (2016) and Mellor et al. (2019)

For the usage of finite groups $\mathcal{D}_{n\mathrm{h}}(\mathrm{M})$ in place of the infinite groups $\mathcal{D}_{\infty\mathrm{h}}(\mathrm{M})$ for building symmetry-adapted basis sets for linear molecules, see Chubb et al. (2018), while artificial symmetries are described in Mellor et al. (2021).

CHAPTER 7

Applications

In this chapter we discuss two common applications of ro-vibrational solutions: calculation of intensities and refinement of PESs. As before, the we assume the finite basis-set representation (FBR) product form for wavefunctions and the sum-of-products form for dipole moment and potential energy functions. We show how the ro-vibrational eigenfunctions corresponding to an *ab initio* PES can be used as basis set in the refinement procedure.

7.1 INTENSITY CALCULATIONS

The ro-vibrational energies are responsible only for part of the spectroscopic information: the line positions. The shape of the spectrum is determined by the line intensities (and line profiles). Computationally, line intensities are probabilities of the ro-vibrational transitions weighted with the corresponding temperature-dependent population of the initial state, e.g. the Boltzmann factor in the case of the local thermal equilibrium (LTE). Here we consider the most common spectroscopic case of the transitions due to electric dipole. The corresponding transition probabilities can be evaluated via matrix elements of the electric dipole moment on the wavefunctions representing the initial and final states. The wavefunctions are the eigenfunctions of the ro-vibrational Hamiltonian, obtained together with the ro-vibrational energies (eigenvalues). The electric dipole moment of a molecule is a three-component vector usually obtained *ab initio* and then parameterised as an analytic function of internal (vibrational) coordinates. In this section we derive general expressions for calculating intensities of electric dipole transitions using ro-vibrational wavefunctions obtained variationally.

7.1.1 General formulation of ro-vibrational intensities

The intensity of an electric dipole transition in absorption or emission depends on two main factors: (a) the number of molecules in the initial state of the transition and (b) the line strength $S(\mathrm{f} \leftarrow \mathrm{i})$, which determines the probability that a molecule in the initial state 'i' will end up in the final state 'f' within unit time.

DOI: 10.1201/9780429154348-7

The line strength of the electric dipole transition between non-degenerate states is given by (Bunker and Jensen, 1998)

$$S(\mathrm{f} \leftarrow \mathrm{i}) = \sum_{A=X,Y,Z} |\langle \Phi_\mathrm{f} | \mu_A | \Phi_\mathrm{i} \rangle|^2 . \tag{7.1}$$

Here μ_A is the component of the molecular dipole moment along the space-fixed axis $A = X$, Y or Z, and $|\Phi_\mathrm{i}\rangle$ and $|\Phi_\mathrm{f}\rangle$ are the complete internal wavefunctions of the initial and final states, respectively. In this book, we follow the Born-Oppenheimer approximation and assume the molecular energies to be independent of nuclear spin (i.e. we neglect the hyperfine structure). The complete internal wavefunctions are therefore given as

$$|\Phi_w\rangle = |\Phi_\mathrm{ns}^{(w)}\rangle \, |\Phi_\mathrm{elec}^{(w)}\rangle \, |\Phi_\mathrm{rv}^{(w)}\rangle, \tag{7.2}$$

where $w = \mathrm{i}$ or f.

Using these wavefunctions for $|\langle \Phi_\mathrm{f} | \mu_A | \Phi_\mathrm{i} \rangle|^2$ from Eq. (7.1), we obtain

$$|\langle \Phi_\mathrm{f} | \mu_A | \Phi_\mathrm{i} \rangle|^2 = \left|\left\langle \Phi_\mathrm{ns}^{(\mathrm{f})} | \Phi_\mathrm{ns}^{(\mathrm{i})} \right\rangle\right|^2 \left|\left\langle \Phi_\mathrm{rv}^{(\mathrm{f})} \, |\bar{\mu}_A| \, \Phi_\mathrm{rv}^{(\mathrm{i})} \right\rangle\right|^2 , \tag{7.3}$$

where we have made use of the fact that the dipole moment component μ_A does not depend on the nuclear spin and introduced the following notation for the electronically averaged electric dipole moment functions:

$$\bar{\mu}_A = \left\langle \Phi_\mathrm{elec}^{(\mathrm{f})} \, |\mu_A| \, \Phi_\mathrm{elec}^{(\mathrm{i})} \right\rangle .$$

The nuclear spin functions in the integral on the left hand side of Eq. (7.3) are orthogonal and normalised:

$$\left|\left\langle \Phi_\mathrm{ns}^{(\mathrm{f})} \Phi_\mathrm{ns}^{(\mathrm{i})} \right\rangle\right|^2 = 1. \tag{7.4}$$

So far we ignored the presence of the nuclear spin under the assumption that the molecular energy does not depend on the nuclear spin state. However, depending on the permutation properties of the ro-vibrational wavefunctions $\Phi_\mathrm{rv}^{(w)}$, not all combinations with $\Phi_\mathrm{ns}^{(w)}$ are allowed. According to the Pauli-principle, the complete internal state for the molecule must obey Fermi-Dirac statistics by permutations of identical fermion nuclei and Bose-Einstein statistics by permutations of identical boson nuclei. Namely, the complete internal wavefunctions have to be invariant to any permutation of identical bosons in the molecule and to any even permutation of identical fermions, but will be changed in sign by an odd permutation of identical fermions. Restrictions imposed by the Pauli-principle on the combinations between the ro-vibrational and nuclear spin parts are directly associated with the permutation (symmetry) properties of $\Phi_\mathrm{rv}^{(w)}$.

Let us assume that for a given initial ro-vibrational state $|\Phi_\mathrm{rv}^{(\mathrm{i})}\rangle$, depending on its symmetry Γ_i, there are g_ns combinations with different nuclear spin functions $|\Phi_\mathrm{ns}^{(\mathrm{i})}\rangle$ available. By the condition in Eq. (7.4), the same g_ns nuclear spin functions can be combined with the final ro-vibrational state $|\Phi_\mathrm{rv}^{(\mathrm{f})}\rangle$ to form allowed complete internal states $|\Phi_\mathrm{f}\rangle$. The quantity g_ns is referred to as the nuclear spin statistical weight factor.

Apart from the g_ns-fold degeneracy of the upper and lower states, they also have m-degeneracy since for a molecule in field-free space, the energy does not depend on m, the projection of the total angular momentum on the Z axis. In order to account for the nuclear spin degeneracy and the m-degeneracy in calculating $S(\mathrm{f} \leftarrow \mathrm{i})$, we generalise Eq. (7.1) to

$$S(\mathrm{f} \leftarrow \mathrm{i}) = g_\mathrm{ns} \sum_{m_\mathrm{f},m_\mathrm{i}} \sum_{A=X,Y,Z} \left| \left\langle \Phi_\mathrm{rv}^{(\mathrm{f})}\, \Phi_\mathrm{elec}^{(\mathrm{f})} \left| \mu_A \right| \Phi_\mathrm{elec}^{(\mathrm{i})}\, \Phi_\mathrm{rv}^{(\mathrm{i})} \right\rangle \right|^2, \tag{7.5}$$

where we have made use of Eq. (7.3) with $|\langle \Phi_\mathrm{ns}^{(\mathrm{f})} | \Phi_\mathrm{ns}^{(\mathrm{i})} \rangle|^2 = 1$.

An absorption line intensity for the transition from the state i with energy E_i, in thermal equilibrium at the temperature T, to the state f with energy E_f is given by (SI units)

$$I(\mathrm{f} \leftarrow \mathrm{i}) \;=\; \frac{8\pi^3 \nu_\mathrm{if}}{(4\pi\epsilon_0)3hc} \frac{e^{-E_\mathrm{i}/kT}}{Q} \left[1 - \exp(-h\nu_\mathrm{fi}/kT)\right] S(\mathrm{f} \leftarrow \mathrm{i}), \tag{7.6}$$

where $\nu = (E_\mathrm{f} - E_\mathrm{i})$ is the line position in Hz (s^{-1}), h is Planck's constant, c is the speed of light in vacuum, k is the Boltzmann constant, ϵ_0 is the permittivity of free space and $S(\mathrm{f} \leftarrow \mathrm{i})$ is the line strength defined in Eq. (7.5). Finally, Q is the partition function defined as

$$Q(T) = \sum_j g_j\, \mathrm{e}^{-E_j/kT}, \tag{7.7}$$

where g_j is the total degeneracy of the ro-vibrational state with energy E_j, which in turn is given by

$$g_j = g_\mathrm{ns}^{(j)} J_j(J_j + 1) \tag{7.8}$$

and the sum runs over all energy levels of the molecule. In Eq. (7.8), J_j is the rotational angular momentum quantum number J of the state j and $g_\mathrm{ns}^{(j)}$ is the nuclear statistical weight or nuclear degeneracy.

The definition of the line strength $S(\mathrm{f} \leftarrow \mathrm{i})$ can be ambiguous depending on the convention to include or exclude the nuclear degeneracy. Here, the line strength $S(\mathrm{f} \leftarrow \mathrm{i})$ takes into account all the degeneracies including the nuclear state degeneracy via the factor g_ns, while some other conventions move this term into the intensity expression. It is therefore common to express the line intensity in terms of the Einstein coefficient A (s^{-1}), which does not have the convention problem and is given by

$$I(\mathrm{f} \leftarrow \mathrm{i}) \;=\; \frac{c^2(2J_\mathrm{i} + 1)}{8\pi\nu_{if}^2 Q}\, e^{-E_\mathrm{i}/kT} \left[1 - e^{-h\nu_\mathrm{fi}/kT}\right] A(\mathrm{f} \leftarrow \mathrm{i}), \tag{7.9}$$

where $hc\,\tilde{\nu}_{\rm fi} = E_{\rm f} - E_{\rm i}$, J' and E' are the angular momentum and energy of the upper state, and the Einstein conference $A({\rm f} \leftarrow {\rm i})$ is given by (using the convention in Eq. (7.5))

$$A({\rm f} \leftarrow {\rm i}) = \frac{16\,\pi^3\,\nu^3}{3\epsilon_0 h\,c^3(2J_{\rm f}+1)} S({\rm f} \leftarrow {\rm i}). \tag{7.10}$$

It is also useful to quote this expression using the CGS space-fixed dipole units as given by

$$A({\rm f} \leftarrow {\rm i}) = \frac{64 \times 10^{-36}\,\pi^4\,\tilde{\nu}_{\rm fi}^3}{3h} S({\rm f} \leftarrow {\rm i}),$$

where h is the Planck constant in the CGS units and the line strength $S({\rm f} \leftarrow {\rm i})$ is in Debye^2.

7.1.2 Calculating the line strength

As discussed above, we describe the rotation of the molecule by means of a molecule-fixed axis system xyz. The orientation of the xyz axis system relative to the XYZ system is defined by the three standard Euler angles (θ, ϕ, χ). Since the molecular dipole moments are commonly defined in the xyz frame, we would like to express the space-fixed dipole moment components (μ_X, μ_Y, μ_Z) given in Eq. (7.5) in terms of the components (μ_x, μ_y, μ_z) along the molecule-fixed axes, which is conveniently performed using the irreducible spherical tensor representation. To this end, we first express the space-fixed dipole moment components using the irreducible tensor representation as

$$\mu_{\rm s}^{(1,\pm 1)} = \frac{1}{\sqrt{2}}(\mp\mu_X - i\,\mu_Y) \text{ and } \mu_{\rm s}^{(1,0)} = \mu_Z \tag{7.11}$$

with analogous expressions for the molecule-fixed components

$$\mu_{\rm m}^{(1,\pm 1)} = \frac{1}{\sqrt{2}}(\mp\mu_x - i\,\mu_y) \text{ and } \mu_{\rm m}^{(1,0)} = \mu_z. \tag{7.12}$$

The transformation between the space-fixed and molecule-fixed components is then given by

$$\mu_{\rm s}^{(1,\sigma)} = \sum_{\sigma'=-1}^{1} [D_{\sigma\sigma'}^{(1)}(\phi,\theta,\chi)]^*\,\mu_{\rm m}^{(1,\sigma')}, \tag{7.13}$$

where $D_{\sigma\sigma'}^{(1)}(\phi,\theta,\chi)$ is an element of a Wigner D-function of rank 1.

The line strength $S({\rm f} \leftarrow {\rm i})$ in Eq. (7.5) in the representation of the irreducible spherical tensors is then given by

$$S({\rm f} \leftarrow {\rm i}) = g_{\rm ns} \sum_{m_f, m_i} \sum_{\sigma=-1}^{1} \left| \left\langle \Psi_{\rm rv}^{\rm (f)} \left| \mu_{\rm s}^{(1,\sigma)} \right| \Psi_{\rm rv}^{\rm (i)} \right\rangle \right|^2. \tag{7.14}$$

The variationally obtained rotation-vibration wavefunctions $|\Psi_{\rm rv}^{(i)}\rangle$ are expansions in terms of the basis function of the form:

$$|\Psi_{\rm rv}^{(i)}\rangle = \sum_{vk} C_{vk}^{(i)}\,|v\rangle|J\,k,\,m\rangle, \tag{7.15}$$

where $C_{vk}^{(i)}$ are the expansion coefficients obtained as eigenvector components in the diagonalisation of the Hamiltonian matrix Eq. (3.5), and $|v\rangle$ is a generic vibrational basis function (see, e.g., Eq. (5.7)) with v used as a short hand notation for all the vibrational quantum numbers v_1, v_2, ..., v_M, vibrational symmetry labels $\Gamma_{\rm vib}$, etc. Substituting $|\Psi_{\rm rv}^{(i)}\rangle$ from Eq. (7.15) into Eq. (7.5) both for the initial and final state wavefunctions, we obtain

$$\begin{aligned} S({\rm f}\leftarrow{\rm i}) &= g_{ns}\sum_{m_{\rm i},m_{\rm f}}\sum_{\sigma=-1}^{1}\Bigg|\sum_{v'k'}\sum_{v''k''} C_{v'k'}^{({\rm f})*}\,C_{v''k''}^{({\rm i})} \\ &\times \sum_{\sigma'=-1}^{1}\left\langle v'\left|\bar{\mu}_{\rm m}^{(1,\sigma')}\right|v''\right\rangle \\ &\times \left\langle J'\,k'\,m_{\rm f}\left|[D_{\sigma\sigma'}^{(1)}(\phi,\theta,\chi)]^*\right|J''\,k''\,m_{\rm i}\right\rangle\Bigg|^2. \end{aligned} \tag{7.16}$$

Here $\bar{\mu}_{\rm m}^{(1,\sigma')}$ is the electronically averaged molecule-fixed dipole moment component which depends on the vibrational coordinates only, whereas the rigid rotor wavefunctions $|J\,k\,m\rangle$ and $[D_{\sigma\sigma'}^{(1)}(\phi,\theta,\chi)]^*$ depend solely on the Euler angles (θ,ϕ,χ). The dipole moment operators

$$\bar{\mu}_{\rm m}^{(1,\sigma')} = \left\langle \Psi_{\rm elec}^{(w)}\left|\mu_{\rm m}^{(1,\sigma')}\right|\Psi_{\rm elec}^{(w)}\right\rangle_{\rm el} \tag{7.17}$$

are assumed to originate from electronic structure calculations as averages over the electronic coordinates. Now using the standard expression and properties of the integrals of $[D_{\sigma\sigma'}^{(1)}(\phi,\theta,\chi)]^*$ over $|J\,k\,m\rangle$, we obtain:

$$\begin{aligned} &S({\rm f}\leftarrow{\rm i}) = g_{ns}\,(2J'+1)\,(2J''+1) \\ &\times\left|\sum_{\sigma=-1,0,1}\sum_{v'k'v''k''} C_{v'k'}^{({\rm f})*}\,C_{v''k''}^{({\rm i})}\,(-1)^{k'}\begin{pmatrix} J'' & 1 & J' \\ k'' & \sigma & -k'\end{pmatrix}\langle v'|\bar{\mu}_{\rm m}^{(1),\sigma}|v''\rangle\right|^2, \end{aligned} \tag{7.18}$$

where

$$\begin{pmatrix} J'' & 1 & J' \\ k'' & \sigma & -k'\end{pmatrix}$$

is the standard $3j$-symbol.

In practice, it is more natural to deal with the Cartesian dipole moment components $\bar{\mu}_{\rm m}^{\alpha}$ ($\alpha = x, y, z$) in place of the spherical tensorial components $\bar{\mu}_{\rm m}^{(1),\sigma}$. Therefore, it often makes sense to transform it back to $\bar{\mu}_{\rm m}^{\alpha}$ in Eq. (7.18)

$$\bar{\mu}_{\rm m}^{(1),\sigma} = \sum_{\alpha=x,y,z} M_{\sigma,\alpha}\bar{\mu}_{\alpha}$$

via the unitary transformation $\boldsymbol{M}$ given by (as deduced from Eq. (7.11))

$$\boldsymbol{M} = \begin{pmatrix} \frac{1}{\sqrt{2}} & \frac{-i}{\sqrt{2}} & 0 \\ \frac{-1}{\sqrt{2}} & \frac{-i}{\sqrt{2}} & 0 \\ 0 & 0 & 1 \end{pmatrix},$$

where $\sigma = -1, 0, 1$ (row index) and $\alpha = x, y, z$ (column index).

Another important consideration is that in practical calculations the rigid rotor wavefunctions are replaced by symmetry-adapted functions $|J, K, \tau_{\rm rot}\rangle$, either in the form of the Wang functions with the well-defined rotational parity $\tau_{\rm rot}$ (see Eq. (6.48)) or, as in the case of the $\mathcal{T}_{\rm d}$-symmetry molecules (see Section 6.6), in the form of a general, symmetry-adapted sum:

$$|J, \lambda, m\rangle = \sum_{k=-J}^{J} T_{\lambda,k}^{(J)} |J, k, m\rangle,$$

where the index λ refers to the rotational symmetry labels $\Gamma_{\rm rot}$ and n as in Eq. (6.69). For the Wang-type rotational basis functions, the $\boldsymbol{T}$-matrix is given in Eqs. (6.63–6.68) and λ refers to the rotational quantum numbers K and $\tau_{\rm rot}$.

Assuming now that the variational eigenfunctions $\Psi_{\rm rv}^{(i)}$ are in the form of Eq. (7.15), the line strength in Eq. (7.18) becomes

$$S({\rm f} \leftarrow {\rm i}) = g_{ns}\,(2J'+1)\,(2J''+1) \Bigg| \sum_{\sigma=-1,0,1} \sum_{\alpha=x,y,z} \sum_{v'\lambda'} \sum_{v''\lambda''} C_{v'\lambda'}^{({\rm f})*} C_{v''\lambda''}^{({\rm i})} \times$$
$$\times T_{\lambda',k'}^{(J')*} T_{\lambda'',k''}^{(J'')} (-1)^{k'} M_{\sigma,\alpha} \begin{pmatrix} J'' & 1 & J' \\ k'' & \sigma & -k' \end{pmatrix} \langle v'|\bar{\mu}_{\alpha}|v''\rangle \Bigg|^2, \qquad (7.19)$$

which is conveniently expressed in terms of the original Cartesian dipole moments in the molecular frame and symmetry-adapted rotational basis functions of the type $|J, \Gamma_{\rm rot}, n, m\rangle$ or $|J, \tau_{\rm rot}, m\rangle$.

Because of $\boldsymbol{M}$, the form given in Eq. (7.19) requires a complex number implementation. It can be shown that for the case of real eigen coefficients, the intensity expression can be presented using real numbers only. For example, for the Wang-type rotational basis set from Eq. (6.48) with the $(-1)^{J+k+\tau_{\rm rot}}$ phase choice, the line strength expression becomes

$$
\begin{aligned}
S(\mathrm{f} \leftarrow \mathrm{i}) &= g_{ns}\,(2J'+1)\,(2J''+1) \\
&\times \Bigg| (i) \sum_{v'K'\tau'_{\mathrm{rot}}} \sum_{v''K''\tau''_{\mathrm{rot}}} C^{(\mathrm{f}*)}_{v'K'\tau'_{\mathrm{rot}}} C^{(\mathrm{i})}_{v''K''\tau''_{\mathrm{rot}}} (-1)^{K''} \\
&\times \Bigg\{ \delta_{K',K''}\delta_{\tau',\tau''}\,(\tau'_{\mathrm{rot}} - \tau''_{\mathrm{rot}}) \begin{pmatrix} J'' & 1 & J' \\ K'' & 0 & -K' \end{pmatrix} \langle v''|\bar{\mu}_z|v''\rangle \\
&+ \delta_{K',K''\pm 1} \begin{pmatrix} J'' & 1 & J' \\ K'' & K'-K'' & -K' \end{pmatrix} \left[\left(\frac{1}{\sqrt{2}} - 1\right) \delta_{K''0}\delta_{K'0} + 1 \right] \\
&\times \left[(K'-K'')(\tau'_{\mathrm{rot}} - \tau''_{\mathrm{rot}})\, \delta_{\tau'_{\mathrm{rot}},\tau''_{\mathrm{rot}}\pm 1} \langle v'|\bar{\mu}_x|v''\rangle \right. \\
&- \left. \delta_{\tau',\tau''} \langle v'|\bar{\mu}_y|v''\rangle \right] \Bigg\} \Bigg|^2 ,
\end{aligned} \tag{7.20}
$$

where elements are real and the imaginary unit i in front of the double sum disappears with the evaluation of the modulus.

7.2 REFINEMENT OF POTENTIAL ENERGY SURFACES

The common situation in high-resolution spectroscopic calculations is that the modern state-of-the-art *ab initio* theory does not provide sufficiently accurate molecular PESs. The *ab initio* PESs are therefore commonly refined by fitting to experimental data, such as energies or line positions. In this section, we explore the methodology for efficient fitting of parameterised potential energy functions (PEFs) as part of the variational calculations.

Let us assume that a PEF $V(\boldsymbol{\xi})$ can be conveniently represented by a *linear* analytically parameterised expression in terms of the internal coordinates ξ_i, e.g.

$$
V = \sum_{ijk\ldots} f_{ijk\ldots} \xi_1^i \xi_2^j \xi_3^k \ldots , \tag{7.21}
$$

where $f_{ijk\ldots}$ are adjustable potential parameters. In turn, the internal coordinates ξ_i may depend (non-linearly) on other structural parameters, e.g. equilibrium constants or Morse constants.

In a refinement of the PES, the potential parameters $f_{ijk\ldots}$ are varied with the goal of minimising the difference between calculated and experimental rovibrational energies defined as the following functional $F(f_{ijk\ldots})$

$$
F(f_{ijk\ldots}) = \sum_{\lambda} \frac{1}{\sigma_\lambda} \left(E^{(\mathrm{exp})}_\lambda - E^{(\mathrm{calc})}_\lambda(\boldsymbol{f}) \right)^2 , \tag{7.22}
$$

where $E^{(\mathrm{exp})}_\lambda$ are experimental ro-vibrational energies with corresponding uncertainties σ_λ and $E^{(\mathrm{calc})}_\lambda(\boldsymbol{f})$ are eigenvalues of the Hamiltonian

$$
\hat{H} = \hat{T} + \sum_{ijk\ldots} f_{ijk\ldots} \xi_1^i \xi_2^j \xi_3^k \ldots
$$

computed using a set of (potential) parameters $\boldsymbol{f} = \{f_{ijk\ldots}\}$. An example of the minimisation procedure based on Newton's minimisation method is detailed below.

Typically, many fitting approaches involve derivatives of the functional $F(f_{ijk\ldots})$ with respect to adjustable parameters $f_{ijk\ldots}$ and require the corresponding first derivatives, such as $\partial E_\lambda^{(J,\Gamma)}/\partial f_{ijk\ldots}$, where $E_\lambda^{(J,\Gamma)}$ are ro-vibrational energies, generated using a given set of the potential parameters $f_{ijk\ldots}$ and λ is a running index to distinguish eigenfunctions with the same values of J and symmetry Γ. The derivatives can be always (but not necessary most efficiently) evaluated numerically using the finite differences as, e.g.

$$\frac{\partial E_\lambda^{(J,\Gamma)}}{\partial f_{ijk\ldots}} = \frac{E_\lambda^{(J,\Gamma)}(f_{ijk\ldots}+h) - E_\lambda^{(J,\Gamma)}(f_{ijk\ldots}-h)}{2h}, \tag{7.23}$$

where h is a small finite difference parameter and the central difference form is used. Here we implicitly assumed the functional form

$$E_\lambda^{(J,\Gamma)} = E_\lambda^{(J,\Gamma)}(f_{ijk\ldots})$$

in the following sense

$$E_\lambda^{(J,\Gamma)} = \langle \Psi_\lambda^{(J,\Gamma)} | \hat{H} | \Psi_\lambda^{(J,\Gamma)} \rangle,$$

where $\Psi_\lambda^{(J,\Gamma)}$ is an eigenfunction of $\hat{H}$, which we assume to depend on $f_{ijk\ldots}$ via the parameterised form of our PEF $V(\boldsymbol{\xi})$ as given by

$$\hat{H} = \hat{T} + V = \hat{T} + \sum_{ijk\ldots} f_{ijk\ldots} \xi_1^i \xi_2^j \xi_3^k \cdots \tag{7.24}$$

with $\hat{T}$ as the kinetic energy operator.

The disadvantage of using the finite differences approach in the refinement is that it requires repeating the eigenvalue calculations multiple times, at least as many times as the number of refining parameters or even twice as many for the central finite differences. A more efficient way to evaluate the derivatives of energies with respect to $f_{ijk\ldots}$ is to use the Hellmann-Feynman theorem. According to Hellmann-Feynman, a derivative of the energies can be represented as an expectation value of the corresponding derivative of the Hamiltonian, which in our case is given by

$$\frac{\partial E_\lambda^{(J,\Gamma)}}{\partial f_{ijk\ldots}} = \langle \Psi_\lambda^{(J,\Gamma)} | \frac{\partial \hat{H}}{\partial f_{ijk\ldots}} | \Psi_\lambda^{(J,\Gamma)} \rangle. \tag{7.25}$$

A linear form of the expansion in Eq. (7.24) in $f_{ijk\ldots}$ helps to simplify the derivative in Eq. (7.25) to

$$\frac{\partial E_\lambda^{(J,\Gamma)}}{\partial f_{ijk\ldots}} = \langle \Psi_\lambda^{(J,\Gamma)} | \xi_1^i \xi_2^j \xi_3^k \cdots | \Psi_\lambda^{(J,\Gamma)} \rangle, \tag{7.26}$$

which does not depend on the adjustable parameters and therefore has to be computed only once for a given set of wavefunctions. It should be noted that $[\xi_1^i \xi_2^j \xi_3^k \ldots]$

is a pure vibrational object; therefore, the corresponding matrix elements are diagonal in the rotational basis used. For instance, assuming the form of $\Psi_\lambda^{(J,\Gamma)}$ as in Eq. (7.15), the ro-vibrational matrix elements in Eq. (7.26) are given by

$$\frac{\partial E_\lambda^{(J,\Gamma)}}{\partial f_{ijk\ldots}} = \sum_{v'v''}\left[\sum_{k'} C_{v'k',\lambda}^{(J,\Gamma)*} C_{v''k',\lambda}^{(J,\Gamma)}\right] \langle v'|\, \xi_1^i \xi_2^j \xi_3^k \ldots \,|v''\rangle.$$

For the variationally computed solutions $E_\lambda^{(J,\Gamma)}$ and $\Psi_\lambda^{(J,\Gamma)}$, the refinement can be facilitated by taking advantage of the '*ab initio*' eigenfunctions (i.e. eigenfunctions that correspond to the 'unrefined' PEF) by using them as basis functions. If we assume that the initial (*ab initio*) PEF $V(\boldsymbol{\xi})$ is a reasonable approximation for the 'true' PEF of the molecule in question, the eigenfunctions of the initial Hamiltonian should provide a good basis set for the refinement and thus significantly reduce the size of the problem. The best basis set is always the one that is closest to the final solution, which for the refinement can be assumed to correspond to the *ab initio* eigenfunctions.

Following this strategy, we factorise the Hamiltonian operator as

$$\hat{H} = \hat{T} + V + \Delta V = \hat{H}_0 + \sum_{ijk\ldots} \Delta f_{ijk\ldots} \{\xi_1^i \xi_2^j \xi_3^k \ldots\}^A, \tag{7.27}$$

where $\hat{H}_0 = T + V$ is the initial, 'unperturbed' Hamiltonian and the effect of the refinement is represented as an additive correction ΔV to V with the refined potential parameters factorised as

$$f_{ijk\ldots} = f_{ijk\ldots}^0 + \Delta f_{ijk\ldots}.$$

The expansion coefficients $\Delta f_{ijk\ldots}$ are the parameters to be refined. Here $\{\xi_1^i \xi_2^j \xi_3^k \ldots\}^A$ correspond to totally symmetric permutations of terms $\xi_1^i \xi_2^j \xi_3^k \ldots$ with A as the totally symmetric representation of the molecular symmetry group, introduced to take advantage of the symmetry properties of the basis functions, if possible.

We first eigen-solve the unperturbed (*ab initio*) Hamiltonian

$$\hat{H}_0 \psi_{0,i}^{(J,\Gamma)} = E_{0,i}^{(J,\Gamma)} \psi_{0,i}^{(J,\Gamma)} \tag{7.28}$$

and use its eigenfunctions $\psi_{0,i}^{(J,\Gamma)}$ as basis functions for the refinement of V. The corresponding matrix elements of $\hat{H}$ in the $\hat{H}_0$ representation are then given by

$$\langle \psi_{0,i}^{(J,\Gamma)} | \hat{H} | \psi_{0,i'}^{(J,\Gamma)} \rangle = E_{0,i}^{(J,\Gamma)} + \sum_{ijk\ldots} \Delta f_{ijk\ldots} \Xi_{i,i'}^{(J,\Gamma)}, \tag{7.29}$$

where

$$\Xi_{i,i'}^{(J,\Gamma)} = \langle \psi_{0,i}^{(J,\Gamma)} | \{\xi_1^i \xi_2^j \xi_3^k \ldots\}^A | \psi_{0,i'}^{(J,\Gamma)} \rangle. \tag{7.30}$$

Assuming that the 'perturbation' ΔV is small, the diagonalisation of Eq. (7.29) can be carried out with a relatively small number of basis functions $\psi_{0,i'}^{(J,\Gamma)}$. Since each expansion term $\{\xi_1^i\xi_2^j\xi_3^k\ldots\}^A$ is totally symmetric in the molecular symmetry group in question, the Hamiltonian matrix in Eq. (7.29) is diagonal both in terms of J and Γ, which significantly reduces the number of the matrix elements in Eq. (7.29) to be evaluated.

Refinement is usually an iterative process, where the current solution n is based on the improvement obtained at the previous step $n-1$. Let us assume $\Psi_\lambda^{(J,\Gamma)}$ is a solution in the H_0-representation at step $n-1$ as given by

$$\Psi_\lambda^{(J,\Gamma)} = \sum_i C_{i,\lambda}^{(J,\Gamma)} \psi_{0,i}^{(J,\Gamma)}. \tag{7.31}$$

At step n, the energy gradients given in Eq. (7.26) can be evaluated using the coefficients $C_{i,\lambda}^{(J,\Gamma)}$ from step $n-1$ as follows:

$$\frac{\partial E_\lambda^{(J,\Gamma)}}{\partial \Delta f_{ijk\ldots}} = \mathbf{C}_i^\dagger\, \boldsymbol{\Xi}^{J\Gamma}\, \mathbf{C}_i, \tag{7.32}$$

where the matrix elements $\boldsymbol{\Xi}^{J\Gamma}$ are from Eq. (7.30), the eigenvectors $\mathbf{C}_i$ contain the coefficients $C_{i,\lambda}^{(J,\Gamma)}$ and the superscript $\dagger$ denotes Hermitian conjugation. It should be also noted that the evaluation of the derivatives in Eq. (7.32) is only required for a limited number of states that are experimentally characterised, which is always significantly smaller than the dimension of the eigenvalue problem.

An important contribution to the energy derivative comes from the so-called zero-point energy (ZPE). ZPE is the lowest (ground) eigenvalue of the system, defined relative to some reference energy, e.g. to the minimum of the PES or to the dissociation threshold. It is useful because only relative values of the ro-vibrational energies (or term values) are usually important. Since $E_{\rm ZPE}$ depends on the potential parameters, it is therefore also affected by the fit, which in turn affects the derivatives of the relative energies $\Delta E_\lambda^{(J,\Gamma)} = E_\lambda^{(J,\Gamma)} - E_{0,0}^{(0,A_1)}$ and has to be taken into account by replacing $E_\lambda^{(J,\Gamma)}$ in Eqs. (7.28–7.32) with $\Delta E_\lambda^{(J,\Gamma)}$. The corresponding derivatives have to be therefore adjusted by subtracting the corresponding ZPE term as follows:

$$\frac{\partial \Delta E_\lambda^{(J,\Gamma)}}{\partial \Delta f_{ijk\ldots}} = \frac{\partial E_\lambda^{(J,\Gamma)}}{\partial \Delta f_{ijk\ldots}} - \frac{\partial E_{\rm ZPE}}{\partial \Delta f_{ijk\ldots}}. \tag{7.33}$$

7.2.1 Fitting using Newton's minimisation method

Here we describe how to use the Newton's method for the minimisation of the functional

$$F(f_{ijk\ldots}) = \sum_\lambda \frac{1}{\sigma_\lambda}\left[E_\lambda^{(\rm exp)} - E_\lambda(\boldsymbol{f})\right]^2 \tag{7.34}$$

(see Eq. (7.22), where we omitted '(calc.)' in $E_\lambda^{(\mathrm{calc})}$ for simplicity) by varying potential parameters $\boldsymbol{f} = \{f_1, f_2, \ldots, f_M\}$ from a molecular PEF $V(\boldsymbol{f})$.

The key feature in the Newton's optimisation technique is the gradients of $F(\boldsymbol{f})$ with respect to $\boldsymbol{f}$

$$\frac{\partial F}{\partial f_i} = -2\sum_{\lambda=1}^{N} \frac{1}{\sigma_\lambda} \left[E_\lambda^{(\mathrm{exp})} - E_\lambda^{(\mathrm{calc})}(\boldsymbol{f})\right] \frac{\partial E_\lambda^{(\mathrm{calc})}(\boldsymbol{f})}{\partial f_i} = 0, \tag{7.35}$$

which are set to vanish to minimise (maximise) $F(\boldsymbol{f})$. Let us now assume that the calculated energies $E_i(\boldsymbol{f})$ have been evaluated on a trial set of parameters $\boldsymbol{f}^{(0)}$. We can now approximate $E_i(\boldsymbol{f})$ by a multi-dimensional linear regression in terms of $\Delta \boldsymbol{f}$ around the trial configuration point $\boldsymbol{f}^{(0)}$ as follows:

$$E_\lambda(\boldsymbol{f}) = E_\lambda(\boldsymbol{f}_0) + \sum_{i=1}^{M} \frac{\partial E_\lambda(\boldsymbol{f}^{(0)})}{\partial f_i} \Delta f_i, \tag{7.36}$$

where

$$\Delta f_i = f_i - f_i^{(0)}$$

and

$$\frac{\partial E_\lambda^{(0)}}{\partial f_i'} \equiv \left.\frac{\partial E_\lambda}{\partial f_i'}\right|_{\boldsymbol{f}=\boldsymbol{f}^{(0)}}$$

is the gradient evaluated at the trial configuration point $\boldsymbol{f}^{(0)}$.

By substituting Eq.(7.36) into Eq. (7.35) after some simplification and rearrangement, we obtain

$$\sum_{i'} \left[\sum_{\lambda=1}^{N} \frac{1}{\sigma_\lambda} \frac{\partial E_\lambda^{(0)}}{\partial f_i} \frac{\partial E_\lambda^{(0)}}{\partial f_{i'}}\right] \Delta f_{i'} = \sum_{\lambda=1}^{N} \frac{1}{\sigma_\lambda} \left[E_\lambda^{(\mathrm{exp})} - E_\lambda^{(0)}\right] \frac{\partial E_\lambda^{(0)}}{\partial f_i}.$$

This is a set of M linear equations for M variables $\Delta f_i'$ in the standard form

$$\boldsymbol{A}\boldsymbol{x} = \boldsymbol{b},$$

for the variables $x_i = \Delta f_i'$, where the elements of the symmetric $M \times M$ matrix $\boldsymbol{A}$ are given by

$$A_{i',i} = \sum_{\lambda=1}^{N} \frac{1}{\sigma_\lambda} \frac{\partial E_\lambda^{(0)}}{\partial f_i} \frac{\partial E_\lambda^{(0)}}{\partial f_i'}$$

and the elements of the vector $\boldsymbol{b}$ are given by

$$b_i = \sum_{\lambda=1}^{N} \frac{1}{\sigma_\lambda} \left[E_\lambda^{(\mathrm{exp})} - E_\lambda^{(0)}\right] \frac{\partial E_\lambda^{(0)}}{\partial f_i}.$$

Assuming that $\boldsymbol{A}$ is invertible, the solution $\boldsymbol{x}$ is efficiently computed using linear algebra libraries. The trial set of the parameters $\boldsymbol{f}^{(0)}$ is then updated in the steepest descent manner to the next iteration as given by

$$f_i^{(1)} = f_i^{(0)} + x_i.$$

The procedure is repeated until the convergence is reached.

7.2.2 Correlated parameters and constrained fit

A very common problem of fitting of the model represented by a large number of parameters required for the description of the full complexity of the spectroscopic model (e.g. potential energy function) is the lack of experimental data. A PEF of a molecule is typically defined by a few hundreds of expansion parameters, but it is very seldom that experimental data cover a sufficiently large number of vibrational excited states to be probed by all the parameters required. Furthermore, not all excitations of the molecule are equally well represented in the experiment. A common situation is that the strongest absorption bands are better characterised. Overpopulated sets of potential parameters lead to an over-fitting, ill-defined or unstable minimisations and result in non-physical PEFs.

A most obvious solution to reduce the number of fitting parameters, which is not always the best, is to constrain some of the potential parameters to their initial, '*ab initio*' values. For example, the lower order expansion parameters are usually more important for reaching the required spectroscopic accuracy. In practice, however, all parameters are correlated, sometimes very strongly. Changing only some of them can lead to a dramatic deformation of the PES.

An alternative approach is to use the so-called 'morphing', when the effect of the refinement is represented as a scaling using a simpler function of the internal coordinates $\boldsymbol{\xi}$:

$$V(\boldsymbol{\xi})^{\text{refined}} = F(\boldsymbol{\xi})^{\text{morph}} \times V(\boldsymbol{\xi})^{\text{ab}},$$

where $F(\boldsymbol{\xi})^{\text{morph}}$ is a parameterised surface representing the effect of the 'morphing'. Providing that $F(\boldsymbol{\xi})^{\text{morph}}$ is a slow varying function of coordinates, 'morphing' can help preserve the original global shape of the *ab initio* PES when introducing sufficient flexibly in-local deformation of PEF.

Another useful alternative allowing one to resolve the correlation of the over-determined parameters is to constrain the PEF through the fit to the original *ab initio* PES. Technically this can be achieved by fitting the PEF (i.e. parameters representing its analytic form) simultaneously to the experimental ro-vibrational energies $E_{\lambda}^{(\text{exp})}$ for different states λ and to the *ab initio* potential energy data points V_n^{ai} at different geometries $\boldsymbol{\xi}_n$. In this case, the functional $\Delta(f_{ijk\ldots})$ in Eq. (7.22) is extended to include the differences between the (reference) *ab initio* PES and the refined PEF as follows:

$$\Delta(f_{ijk\ldots}) = \sum_{\lambda} \frac{1}{\sigma_{\lambda}^{(\text{exp})}} \left(E_{\lambda}^{(\text{exp})} - E_{\lambda}(\boldsymbol{f}) \right)^2 + \omega \sum_{n} \frac{1}{\sigma_{n}^{(\text{ai})}} \left(V_n^{(\text{ai})} - V(\boldsymbol{f}, \xi_n) \right)^2,$$

where $V_{\lambda}^{\text{calc}}(\boldsymbol{f})$ are the corresponding values of potential energy computed at the geometry $\boldsymbol{\xi}_n$ and defined using the same set of potential parameters $\boldsymbol{f}$, ω is a normalised weight factor representing the relative importance of the two sets of data and $\sigma_n^{(\text{ai})}$ is the uncertainty associated with the *ab initio* potential energy value $V_n^{(\text{ai})}$. The required derivatives of the eigenvalues are obtained via Eqs. (7.23) or (7.26), while the derivatives of PEF are simply

$$\frac{\partial V(\boldsymbol{\xi})}{\partial f_{ijk\ldots}} = \xi_1^i \xi_2^j \xi_3^k \ldots$$

A common artefact of empirical adjustments is that different imperfections of the model such as basis set incompleteness or approximations involved can also affect the refined PEF. As a result, these imperfections are effectively absorbed by the 'improved' PEF, thus making it a rather effective object that is able to reproduce the experimental energies with the accuracy achieved only with the same imperfect model used in the refinements. The *ab initio* constraint can provide a measure for the deformation of PEF introduced by the fit as a difference with the *ab initio* data. Controlling the fitting shape can be especially important when the over-fitting is difficult to avoid. Moreover, since lower fitting residuals defined by $\Delta(f_{ijk\ldots})$ do not necessarily mean improvement of the PEF, the deviation from the first principles data is the only objective measure of the shape of the refined PEF.

MATERIALS USED

The description of the intensity methodology is largely based on Bunker and Jensen (1998) and Yurchenko et al. (2005b).

The description of the PES refinement and how to replace the $J = 0$ energies with more accurate experimental values follows Yurchenko et al. (2011, 2020).

The definition of the line strength by Bernath (2015) does not contain the nuclear degeneracy, while the definition by Bunker and Jensen (1998) does. In this book, we use the latter convention.

Description of the *ab initio* constraints via a simultaneous fit to experimental and *ab initio* energies is from Yurchenko et al. (2003).

For the morphing fitting procedure, see, e.g., Meuwly and Hutson (1999) and Yurchenko et al. (2016).

MATERIALS NOT USED BUT WORTH MENTIONING

A generalised method of computing matrix elements of an arbitrary electric tensor $\boldsymbol{T}_\Omega$ of rank Ω including dipole moment, polarisability, hyperpolarisabilities, etc., on variationally obtained ro-vibrational wavefunctions is described in Owens and Yachmenev (2018).

Marquardt's optimisation procedure (Marquardt, 1963) and linear search optimisation method is by Armijo (1966).

An efficient approach to dynamically redefine the fitting weights (experimental uncertainties) is Watson's robust fitting (Watson, 2003).

Bibliography

M. Abramowitz and I. A. Stegun. *Handbook of Mathematical Functions with Formulas, Graphs, and Mathematical Tables.* Dover, New York, Ninth Dover printing, tenth GPO printing edition, 1964.

A. F. Al-Refaie, O. L. Polyansky, R. I. Ovsyannikov, J. Tennyson, and S. N. Yurchenko. ExoMol line lists - XV. A new hot line list for hydrogen peroxide. *Mon. Not. R. Astron. Soc.*, 461:1012–1022, 2016.

O. Alvarez-Bajo, R. Lemus, M. Carvajal, and F. Perez-Bernal. Equivalent rotations associated with the permutation inversion group revisited: Symmetry projection of the rovibrational functions of methane. *Mol. Phys.*, 109:797–812, 2011.

V. Aquilanti and S. Cavalli. Coordinates for molecular dynamics: Orthogonal local systems. *J. Chem. Phys.*, 85:1355–1361, 1986.

L. Armijo. Minimization of functions having Lipschitz continuous first partial derivatives. *Pacific J. Math.*, 16:1–3, 1966.

G. Avila and T. Carrington. Nonproduct quadrature grids for solving the vibrational Schrödinger equation. *J. Chem. Phys.*, 131:174103, 2009.

G. Avila and T. Carrington, Jr. A multi-dimensional Smolyak collocation method in curvilinear coordinates for computing vibrational spectra. *J. Chem. Phys.*, 143:214108, 2015.

A. A. A. Azzam, J. Tennyson, S. N. Yurchenko, and O. V. Naumenko. ExoMol molecular line lists - XVI. The rotation-vibration spectrum of hot H_2S. *Mon. Not. R. Astron. Soc.*, 460(4):4063–4074, 2016.

Z. Bačić and J. C. Light. Theoretical methods for rovibrational states of floppy molecules. *Ann. Rev. Phys. Chem.*, 40:469–498, 1989.

M. H. Beck, A. Jäckle, G. A. Worth, and H. D. Meyer. The multiconfiguration time-dependent Hartree (MCTDH) method: A highly efficient algorithm for propagating wavepackets. *Phys. Rep.*, 324:1–105, 2000.

V. A. Benderskii, E. V. Vetoshkin, I. S. Irgibaeva, and H. P. Trommsdorff. Six-dimensional tunneling dynamics of internal rotation in hydrogen peroxide molecule and its isotopomers. *Russ. Chem. Bull.*, 50:366–375, 2001.

P. F. Bernath. *Spectra of Atoms and Molecules.* Oxford University Press, 3rd edition, 2015.

A. Blondel and M. Karplus. New formulation for derivatives of torsion angles and improper torsion angles in molecular mechanics: Elimination of singularities. *J. Comput. Chem.*, 17(9):1132–1141, 1996.

I. Bohaček, D. Papoušek, Š. Pick, and V. Špirko. Eigenvalue problem for the operator $h = -\frac{1}{2}\mathrm{d}^2/\mathrm{d}\rho^2 + \frac{1}{2}m^2\rho^2 + g/\rho^2$ and the correlation between the energy levels of linear and bent molecules. *Chem. Phys. Lett.*, 42(2):395 – 398, 1976. ISSN 0009-2614.

V. Boudon, J.-P. Champion, T. Gabard, M. Loëte, F. Michelot, G. Pierre, M. Rotger, C. Wenger, and M. Rey. Symmetry-adapted tensorial formalism to model rovibrational and rovibronic spectra of molecules pertaining to various point groups. *J. Mol. Spectrosc.*, 228:620–634, 2004.

M. J. Bramley and T. Carrington. A general discrete variable method to calculate vibrational energy levels of three- and four-atom molecules. *J. Chem. Phys.*, 99:8519–8541, 1993.

M. J. Bramley, W. H. Green, and N. C. Handy. Vibration-rotation coordinates and kinetic-energy operators for polyatomic-molecules. *Mol. Phys.*, 73: 1183–1208, 1991.

J. Brown and T. Carrington. Using an expanding nondirect product harmonic basis with an iterative eigensolver to compute vibrational energy levels with as many as seven atoms. *J. Chem. Phys.*, 145:144104, 2016.

P. R. Bunker and P. Jensen. *Molecular Symmetry and Spectroscopy.* NRC Research Press, Ottawa, 2nd edition, 1998.

P. R. Bunker and P. Jensen. Spherical top molecules and the molecular symmetry group. *Mol. Phys.*, 97:255–264, 1999.

T. Carrington. Iterative methods for computing vibrational spectra. *Mathematics*, 6(1), 13, 2018.

T. Carrington. Using collocation to study the vibrational dynamics of molecules. *Spectra Chimica Acta A*, 248:119158, 2021.

S. Carter and N. Handy. The variational method for the calculation of rovibrational energy levels. *Comput. Phys. Rep.*, 5:117–171, 1986.

S. Carter and N. C. Handy. A variational method for the calculation of vibrational levels of any triatomic molecule. *Mol. Phys.*, 47:1445–1455, 1982.

S. Carter and N. C. Handy. The vibrational levels of C_2H_2 using an internal coordinate vibrational Hamiltonian. *Mol. Phys.*, 53:1033–1039, 1984.

S. Carter and W. Meyer. A variational method for the calculation of vibrational energy levels of triatomic molecules using a Hamiltonian in hyperspherical coordinates. *J. Chem. Phys.*, 93:8902–8914, 1990.

S. Carter and W. Meyer. A variational method for the calculation of rovibrational energy levels of triatomic molecules using a Hamiltonian in hyperspherical coordinates: Applications to H_3^+ and Na_3^+. *J. Chem. Phys.*, 100 (3):2104–2117, 1994.

S. Carter, N. Handy, and B. Sutcliffe. A variational method for the calculation of rovibrational levels of any triatomic molecule. *Mol. Phys.*, 49:745–748, 1983.

S. Carter, N. C. Handy, and J. Demaison. The rotational levels of the ground vibrational state of formaldehyde. *Mol. Phys.*, 90:729–737, 1997.

S. Carter, A. R. Sharma, J. M. Bowman, P. Rosmus, and R. Tarroni. Calculations of rovibrational energies and dipole transition intensities for polyatomic molecules using MULTIMODE. *J. Chem. Phys.*, 131:224106, 2009.

J.-P. Champion, M. Loéte, and G. Pierre. Spherical top spectra. In K. N. Rao and A. Weber, editors, *Spectroscopy of the Earth's Atmosphere and Interstellar Medium*, pages 339–422. Academic Press, San Diego, CA, 1992. ISBN 978-0-12-580645-9. doi: 10.1016/B978-0-12-580645-9.50009-5.

X. Chapuisat and C. Iung. Vector parametrization of the N-body problem in quantum-mechanics - polyspherical coordinates. *Phys. Rev. A*, 45(9): 6217–6235, 1992.

X. Chapuisat, A. Nauts, and J.-P. Brunet. Exact quantum molecular hamiltonians. *Mol. Phys.*, 72(1):1–31, 1991.

J.-Q. Chen, M.-J. Gao, and G.-Q. Ma. The representation group and its application to space groups. *Rev. Mod. Phys.*, 57:211–278, 1985.

K. L. Chubb, P. Jensen, and S. N. Yurchenko. Symmetry adaptation of the rotation-vibration theory for linear molecules. *Symmetry*, 10:137, 2018.

K. L. Chubb, A. Yachmenev, J. Tennyson, and S. N. Yurchenko. Treating linear molecule HCCH in calculations of rotation-vibration spectra. *J. Chem. Phys.*, 149:014101, July 2018.

K. L. Chubb, J. Tennyson, and S. N. Yurchenko. ExoMol molecular line lists - XXXVII: Spectra of acetylene. *Mon. Not. R. Astron. Soc.*, 493:1531–1545, 2020.

J. W. Cooley. An improved eigenvalue corrector formula for solving the Schrödinger equation for central fields. *Math. Comp.*, 15:363–374, 1961.

A. G. Császár, C. Fabri, T. Szidarovszky, E. Mátyus, T. Furtenbacher, and G. Czakó. The fourth age of quantum chemistry: Molecules in motion. *Phys. Chem. Chem. Phys.*, 14:1085–1106, 2012.

A. Y. Dymarsky and K. N. Kudin. Computation of the pseudorotation matrix to satisfy the eckart axis conditions. *J. Chem. Phys.*, 122(12):124103, 2005.

C. Eckart. Some studies concerning rotating axes and polyatomic molecules. *Phys. Rev.*, 47:552–558, 1935.

S. Erfort, M. Tschöpe, and G. Rauhut. Toward a fully automated calculation of rovibrational infrared intensities for semi-rigid polyatomic molecules. *J. Chem. Phys.*, 152:244104, 2020.

C. Fabri, T. Furtenbacher, and A. G. Császár. A hybrid variation-perturbation nuclear motion algorithm. *Mol. Phys.*, 112:2462–2467, 2014.

C. Fábri, E. Mátyus, and A. G. Császár. Numerically constructed internal-coordinate Hamiltonian with Eckart embedding and its application for the inversion tunneling of ammonia. *Spectra Chimica Acta A*, 119:84–89, 2014.

C. Fabri, E. Matyus, and A. G. Császár. Rotating full- and reduced-dimensional quantum chemical models of molecules. *J. Chem. Phys.*, 134: 074105, 2011.

F. Gatti and C. Iung. Exact and constrained kinetic energy operators for polyatomic molecules: The polyspherical approach. *Phys. Rep.*, 484:1–69, 2009.

F. Gatti and A. Nauts. Vector parametrization, partial angular momenta and unusual commutation relations in physics. *Chem. Phys.*, 295(2):167–174, 2003. ISSN 0301-0104.

F. Gatti, C. Iung, M. Menou, and X. Chapuisat. Vector parametrization of the N-atom problem in quantum mechanics. II. Coupled-angular-momentum spectral representations for four-atom systems. *J. Chem. Phys.*, 108(21): 8821–8829, 1998.

F. Gatti, C. Muñoz, and C. Iung. A general expression of the exact kinetic energy operator in polyspherical coordinates. *J. Chem. Phys.*, 114(19): 8275–8281, 2001.

L. Halonen. Internal coordinate Hamiltonian model for Fermi resonances and local modes in methane. *J. Chem. Phys.*, 106:831–845, 1997.

N. C. Handy. The derivation of vibration-rotation kinetic-energy operators, in internal coordinates. *Mol. Phys.*, 61:207–223, 1987.

N. C. Handy and S. Carter. Large vibrational variational calculations using 'multimode' and an iterative diagonalization technique. *Mol. Phys.*, 102: 2201–2205, 2004.

N. C. Handy, S. Carter, and S. M. Colwell. The vibrational energy levels of ammonia. *Mol. Phys.*, 96(4):477–491, 1999.

J. Hougen. *Methane Symmetry Operations* . NIST, Gaithersburg, MD., 2001. Version 1.0 [Online]. Available: `http://physics.nist.gov/Methane`.

J. T. Hougen. Rotational energy levels of a linear triatomic molecule in a $^2\Pi$ electronic state. *J. Chem. Phys.*, 36:519–534, 1962a.

J. T. Hougen. Classification of rotational energy levels for symmetric-top molecules. *J. Chem. Phys.*, 37:1433–1441, 1962b.

J. T. Hougen. Classification of rotational energy levels II. *J. Chem. Phys.*, 39(2):358–365, 1963.

J. T. Hougen. A group-theoretical treatment of electronic, vibrational, torsional, and rotational motions nn the dimethylacetylene molecule. *Can. J. Phys.*, 42:1920–1937, 1964.

J. T. Hougen, P. R. Bunker, and J. W. C. Johns. Vibration-rotation problem in triatomic molecules allowing for a large-amplitude bending vibration. *J. Mol. Spectrosc.*, 34:136–172, 1970.

B. J. Howard and R. E. Moss. The molecular hamiltonian. *Mol. Phys.*, 20: 147–159, 1971.

X. Huang, R. S. Freedman, S. A. Tashkun, D. W. Schwenke, and T. J. Lee. Semi-empirical $^{12}C^{16}O_2$ IR line lists for simulations up to 1500 K and 20,000 cm^{-1}. *J. Quant. Spectrosc. Radiat. Transf.*, 130:134–146, 2013.

P. Jensen. A new morse oscillator-rigid bender internal dynamics (morbid) hamiltonian for triatomic-molecules. *J. Mol. Spectrosc.*, 128:478–501, 1988.

D. W. Jepsen and J. O. Hirschfelder. Set of co-ordinate systems which diagonalize the kinetic energy of relative motion. *Proc. Nat. Acad. Sci.*, 45: 249–256, 1959.

B. R. Johnson. The quantum dynamics of three particles in hyperspherical coordinates. *J. Chem. Phys.*, 79(4):1916–1925, 1983.

B. R. Johnson and W. P. Reinhardt. Adiabatic separations of stretching and bending vibrations: Application to H_2O. *J. Chem. Phys.*, 85(8):4538–4556, 1986.

B. Jordanov and W. Orville-Thomas. A computational method for symmetry factorization without using symmetry operators. *J. Molec. Struct. (THEOCHEM)*, 76(4):323 – 327, 1981. ISSN 0166-1280.

I. Kozin, M. M. Law, J. Tennyson, and J. M. Hutson. New vibration-rotation code for tetraatomic molecules exhibiting wide-amplitude motion: Wavr4. *Comput. Phys. Commun.*, 163:117–131, 2004.

I. N. Kozin, M. M. Law, J. M. Hutson, and J. Tennyson. Calculating energy levels of isomerizing tetra-atomic molecules. I. The rovibrational bound states of Ar_2HF. *J. Chem. Phys.*, 118(11):4896–4904, 2003.

B. Kuhn, T. R. Rizzo, D. Luckhaus, M. Quack, and M. A. Suhm. A new six-dimensional analytical potential up to chemically significant energies for the electronic ground state of hydrogen peroxide. *J. Chem. Phys.*, 111: 2565–2587, 1999.

D. Lauvergnat and A. Nauts. Exact numerical computation of a kinetic energy operator in curvilinear coordinates. *J. Chem. Phys.*, 116:8560–8570, 2002.

D. Lauvergnat, J. M. Luis, B. Kirtman, H. Reis, and A. Nauts. Numerical and exact kinetic energy operator using Eckart conditions with one or several reference geometries: Application to HONO. *J. Chem. Phys.*, 144:084116, 2016.

C. Leforestier, A. Viel, F. Gatti, C. Muñoz, and C. Iung. The Jacobi-Wilson method: A new approach to the description of polyatomic molecules. *J. Chem. Phys.*, 114(5):2099–2105, 2001.

R. Lemus. A general method to obtain vibrational symmetry adapted bases in a local scheme. *Mol. Phys.*, 101:2511–2528, 2003.

R. Lemus. Quantum numbers and the eigenfunction approach to obtain symmetry adapted functions for discrete symmetries. *Symmetry*, 4:667, 2012.

C. Léonard, N. C. Handy, S. Carter, and J. M. Bowman. The vibrational levels of ammonia. *Spectrochimica Acta A*, 58:825–838, 2002.

J. C. Light and T. Carrington Jr. *Discrete-Variable Representations and their Utilization*, pages 263–310. John Wiley & Sons, Ltd, 2000. https://doi.org/10.1002/9780470141731.ch4.

J. C. Light, I. P. Hamilton, and J. V. Lill. Generalized discrete variable approximation in quantum mechanics. *J. Chem. Phys.*, 82:1400–1409, 1985.

R. G. Littlejohn and M. Reinsch. Gauge fields in the separation of rotations andinternal motions in the n-body problem. *Rev. Mod. Phys.*, 69:213–276, Jan 1997.

H. Longuet-Higgins. The symmetry groups of non-rigid molecules. *Mol. Phys.*, 6:445–460, 1963.

J. D. Louck and H. W. Galbraith. Eckart vectors, eckart frames, and polyatomic-molecules. *Rev. Mod. Phys.*, 48:69–106, 1976.

T. J. Lukka. A simple method for the derivation of exact quantum–mechanical vibration–rotation hamiltonians in terms of internal coordinates. *J. Chem. Phys.*, 102(10):3945–3955, 1995.

J. Makarewicz and A. Skalozub. Rovibrational molecular Hamiltonian in mixed bond-angle and umbrella-Like coordinates. *J. Phys. Chem. A*, 111: 7860–7869, 2007.

U. Manthe. A multilayer multiconfigurational time-dependent Hartree approach for quantum dynamics on general potential energy surfaces. *J. Chem. Phys.*, 128:164116, 2008.

D. W. Marquardt. An algorithm for least-squares estimation of nonlinear parameters. *J. Soc. Ind. Appl. Math.*, 11:431–441, 1963.

T. Mathea, T. Petrenko, and G. Rauhut. VCI calculations based on canonical and localized normal coordinates for Non-Abelian molecules: Accurate assignment of the vibrational overtones of allene. *J. Phys. Chem. A*, 125: 990–998, 2021.

E. Mátyus, G. Czakó, B. T. Sutcliffe, and A. G. Császár. Vibrational energy levels with arbitrary potentials using the eckart-watson hamiltonians and the discrete variable representation. *J. Chem. Phys.*, 127:084102, 2007.

E. Mátyus, G. Czakó, and A. G. Császár. Toward black-box-type full- and reduced-dimensional variational (ro)vibrational computations. *J. Chem. Phys.*, 130:134112, 2009a.

E. Mátyus, G. Czakó, and A. G. Császár. Toward black-box-type full- and reduced-dimensional variational (ro)vibrational computations. *J. Chem. Phys.*, 130:134112, 2009b.

E. Mátyus, C. Fabri, T. Szidarovszky, G. Czako, W. D. Allen, and A. G. Császár. Assigning quantum labels to variationally computed rotational-vibrational eigenstates of polyatomic molecules. *J. Chem. Phys.*, 133: 034113, 2010.

T. M. Mellor, S. N. Yurchenko, B. P. Mant, and P. Jensen. Transformation properties under the operations of the molecular symmetry groups G36 and G36(EM) of ethane H_3CCH_3. *Symmetry*, 11(862), 2019.

T. M. Mellor, S. N. Yurchenko, and P. Jensen. Artificial symmetries for calculating vibrational energies of linear molecules. *Symmetry*, 13: 13, 2021.

A. Messiah. *Quantum Mechanics*. North-Holland, Amsterdam, 1961.

M. Meuwly and J. M. Hutson. Morphing ab initio potentials: A systematic study of Ne-HF. *J. Chem. Phys.*, 110:8338–8347, 1999.

R. Meyer and H. H. Günthard. General internal motion of molecules, classical and quantum-mechanical Hamiltonian. *J. Chem. Phys.*, 49(4):1510–1520, 1968.

M. Mladenovic. Rovibrational Hamiltonians for general polyatomic molecules in spherical polar parametrization. I. Orthogonal representations. *J. Chem. Phys.*, 112:1070–1081, 2000.

M. Mladenovic. Discrete variable approaches to tetratomic molecules Part I: DVR(6) and DVR(3)+DGB methods. *Spectra Chimica Acta A*, 58:795–807, 2002.

M. Neff and G. Rauhut. Toward large scale vibrational configuration interaction calculations. *J. Chem. Phys.*, 131:124129, 2009.

H. H. Nielsen. The vibration-rotation energies of molecules. *Rev. Mod. Phys.*, 23:90–136, 1951.

A. Nikitin, J. Champion, and V. Tyuterev. Improved algorithms for the modeling of vibrational polyads of polyatomic molecules: Application to T_d, O_h, and C_{3v} molecules. *J. Mol. Spectrosc.*, 182(1):72 – 84, 1997.

A. V. Nikitin, M. Rey, and V. G. Tyuterev. An efficient method for energy levels calculation using full symmetry and exact kinetic energy operator: Tetrahedral molecules. *J. Chem. Phys.*, 142:094118, 2015.

B. V. Noumerov. A method of extrapolation of perturbations. *Mon. Not. R. Astron. Soc.*, 84:592–602, 1924.

R. I. Ovsyannikov, V. V. Melnikov, W. Thiel, P. Jensen, O. Baum, T. F. Giesen, and S. N. Yurchenko. Theoretical rotation-torsion energies of HSOH. *J. Chem. Phys.*, 129:154314, 2008.

A. Owens and A. Yachmenev. Richmol: A general variational approach for rovibrational molecular dynamics in external electric fields. *J. Chem. Phys.*, 148(12):124102, 2018.

D. Papoušek and M. R. Aliev. *Molecular Vibrational-Rotational Spectra.* Elsevier, Amsterdam, 1982.

W. Pauli. On the connexion between the completion of electron groups in an atom with the complex structure of spectra. *Zeitschrift für Physik*, 31: 765, 1925.

W. Pauli. The connection between spin and statistics. *Phys. Rev.*, 58:716–722, 1940.

J. Pesonen. Eckart frame vibration-rotation Hamiltonians: Contravariant metric tensor. *J. Chem. Phys.*, 140(7):074101, 2014.

J. Pesonen, A. Miani, and L. Halonen. New inversion coordinate for ammonia: Application to a CCSD(T) bidimensional potential energy surface. *J. Chem. Phys.*, 115:1243–1250, 2001.

B. W. Pfennig. *Principles of Inorganic Chemistry.* Wiley, Hoboken, New Jersey, 2015.

B. Podolsky. Quantum-mechanically correct form of Hamiltonian function for conservative systems. *Phys. Rev.*, 32:0812–0816, 1928.

S. R. Polo. Matrices D^{-1} and G^{-1} in the theory of molecular vibrations. *J. Chem. Phys.*, 24(6):1133–1138, 1956.

W. H. Press, S. A. Teukolsky, W. T. Vetterling, and B. P. Flannery. *Numerical Recipes – The Art of Scientific Computing.* Cambridge University Press, 3rd edition, 2007.

A. E. Protasevich and A. V. Nikitin. Kinetic energy operator of linear symmetric molecules of the A2B2 type in polyspherical orthogonal coordinates. *Atmos. Ocean. Opt.*, 35:14–18, 2022.

R. Radau. Sur une transformation des équations différentielles de la dynamique. *Annales scientifiques de l'École Normale Supérieure*, 1e série, 5: 311–375, 1868.

M. Rey. Group-theoretical formulation of an Eckart-frame kinetic energy operator in curvilinear coordinates for polyatomic molecules. *J. Chem. Phys.*, 151:024101, 2019.

M. Rey. Novel methodology for systematically constructing global effective models from ab initio-based surfaces: A new insight into high-resolution molecular spectra analysis. *J. Chem. Phys.*, 156:224103, 2022.

M. Rey, A. Nikitin, and V. Tyuterev. Ab initio ro-vibrational hamiltonian in irreducible tensor formalism: A method for computing energy levels from potential energy surfaces for symmetric-top molecules. *Mol. Phys.*, 108 (16):2121–2135, 2010.

M. Rey, A. V. Nikitin, and V. G. Tyuterev. Complete nuclear motion Hamiltonian in the irreducible normal mode tensor operator formalism for the methane molecule. *J. Chem. Phys.*, 136:244106, 2012.

K. Sadri, D. Lauvergnat, F. Gatti, and H.-D. Meyer. Rovibrational spectroscopy using a kinetic energy operator in Eckart frame and the multi-configuration time-dependent Hartree (MCTDH) approach. *J. Chem. Phys.*, 141:114101, 2014.

J. Sarka and B. Poirier. Hitting the Trifecta: How to simultaneously push the limits of Schrödinger solution with respect to system size, convergence accuracy, and number of computed states. *J. Chem. Theory Comput.*, 17: 7732–7744, 2021.

J. Sárka, B. Poirier, V. Szalay, and A. G. Császár. On neglecting Coriolis and related couplings in first-principles rovibrational spectroscopy: Considerations of symmetry, accuracy, and simplicity. *Sci. Rep.*, 10:4872, 2020.

J. Sarka, B. Poirier, V. Szalay, and A. G. Császár. On neglecting Coriolis and related couplings in first-principles rovibrational spectroscopy: Considerations of symmetry, accuracy, and simplicity. II. Case studies for H2O isotopologues, H_3^+, O_3, and NH_3. *Spectra Chimica Acta A*, 250:119164, 2021.

A. Sayvetz. The kinetic energy of polyatomic molecules. *J. Chem. Phys.*, 7: 383–389, 1939.

B. Schröder. Variational rovibrational calculations for tetra atomic linear molecules using Watson's isomorphic Hamiltonian, I: The C8v4 approach. *J. Mol. Spectrosc.*, 385:111613, 2022.

D. W. Schwenke. Variational calculations of rovibrational energy levels and transition intensities for tetratomic molecules. *J. Phys. Chem.*, 100:2867–2884, 1996.

D. W. Schwenke. New rovibrational kinetic energy operators using polyspherical coordinates for polyatomic molecules. *J. Chem. Phys.*, 118:10431–10438, 2003.

G. O. Sørensen. A new approach to the hamiltonian of nonrigid molecules. In M. J. S. D. et al., editor, *Large Amplitude Motion in Molecules II*, volume 82 of *Topics in Current Chemistry*, pages 97–175. Springer Berlin Heidelberg, Heidelberg, 1979.

C. Sousa-Silva, S. N. Yurchenko, and J. Tennyson. A computed room temperature line list for phosphine. *J. Mol. Spectrosc.*, 288:28–37, JUN 2013.

C. Sousa-Silva, A. F. Al-Refaie, J. Tennyson, and S. N. Yurchenko. ExoMol line lists - VII. The rotation-vibration spectrum of phosphine up to 1500 K. *Mon. Not. R. Astron. Soc.*, 446:2337–2347, 2015.

C. Sousa-Silva, J. Tennyson, and S. N. Yurchenko. Communication: Tunnelling splitting in the phosphine molecule. *J. Chem. Phys.*, 145, 2016.

V. Špirko. Vibrational anharmonicity and the inversion potential function of NH_3. *J. Mol. Spectrosc.*, 101(1):30 – 47, 1983.

H. L. Strauss and H. M. Pickett. Conformational structure, energy, and inversion rates of cyclohexane and some related oxanes. *J. Am. Chem. Soc.*, 92(25):7281–7290, 1970.

B. T. Sutcliffe and J. Tennyson. A generalised approach to the calculation of ro-vibrational spectra of triatomic molecules. *Mol. Phys.*, 58:1053–1066, 1986.

B. T. Sutcliffe and J. Tennyson. A general treatment of vibration-rotation coordinates for triatomic molecules. *Intern. J. Quantum Chem.*, 39:183 – 196, 1991.

V. Szalay. Eckart-Sayvetz conditions revisited. *J. Chem. Phys.*, 140:234107, 2014.

V. Szalay. Aspects of the Eckart frame ro-vibrational kinetic energy operator. *J. Chem. Phys.*, 143:064104, 2015a.

V. Szalay. Understanding nuclear motions in molecules: Derivation of Eckart frame ro-vibrational Hamiltonian operators via a gateway Hamiltonian operator. *J. Chem. Phys.*, 142:174107, 2015b.

V. Szalay. Eckart ro-vibrational hamiltonians via the gateway hamilton operator: Theory and practice. *J. Chem. Phys.*, 146(12):124107, 2017.

T. Szidarovszky, C. Fabri, and A. G. Császár. The role of axis embedding on rigid rotor decomposition analysis of variational rovibrational wave functions. *J. Chem. Phys.*, 136:174112, 2012.

J. Tennyson. The calculation of vibration-rotation energies of triatomic molecules using scattering coordinates. *Comput. Phys. Rep.*, 4:1–36, 1986.

J. Tennyson and B. T. Sutcliffe. Variationally exact ro-vibrational levels of the floppy CH_2^+ molecule. *J. Mol. Spectrosc.*, 101(1):71–82, 1983.

J. Tennyson, M. A. Kostin, P. Barletta, G. J. Harris, O. L. Polyansky, J. Ramanlal, and N. F. Zobov. DVR3D: A program suite for the calculation of rotation-vibration spectra of triatomic molecules. *Comput. Phys. Commun.*, 163:85–116, 2004.

J. Tennyson, M. J. Barber, and R. E. A. Kelly. An adiabatic model for calculating overtone spectra of dimers such as $(H_2O)_2$. *Phil. Trans. Royal Soc. London A*, 370:2656–2674, 2012.

P. S. Thomas and T. Carrington. Using nested contractions and a hierarchical tensor format to compute vibrational spectra of molecules with seven atoms. *J. Phys. Chem. A*, 119:13074–13091, 2015.

R. C. van Schaik, H. J. Berendsen, A. E. Torda, and W. F. van Gunsteren. A structure refinement method based on molecular dynamics in four spatial dimensions. *J. Mol. Biol.*, 234(3):751–762, 1993. ISSN 0022-2836.

D. A. Varshalovich, A. N. Moskalev, and V. K. Khersonskii. *Quantum Theory of Angular Momentum.* World Scientific, Singapor, 1988.

F. Wang, F. McCourt, and E. von Nagy-Felsobuki. An Eckart-Watson Hamiltonian for linear molecules in the rectilinear displacement w-coordinates and an application to HCN. *J. Molec. Struct. (THEOCHEM)*, 497:227–240, 2000. ISSN 0166-1280.

H. Wang. Multilayer multiconfiguration time-dependent Hartree Theory. *J. Phys. Chem. A*, 119:7951–7965, 2015.

S. C. Wang. On the asymmetrical top in quantum mechanics. *Phys. Rev.*, 34: 243–252, 1929.

X.-G. Wang and T. Carrington. A symmetry-adapted Lanczos method for calculating energy levels with different symmetries from a single set of iterations. *J. Chem. Phys.*, 114(4):1473–1477, 2001.

X.-G. Wang and T. Carrington, Jr. Computing rovibrational levels of methane with curvilinear internal vibrational coordinates and an Eckart frame. *J. Chem. Phys.*, 138:104106, 2013.

J. K. Watson. Simplification of the molecular vibration-rotation hamiltonian. *Mol. Phys.*, 15:479–490, 1968.

J. K. Watson. The vibration-rotation hamiltonian of linear molecules. *Mol. Phys.*, 19:465–487, 1970.

J. K. Watson. The molecular vibration-rotation kinetic–energy operator for general internal coordinates. *J. Mol. Spectrosc.*, 228:645 – 658, 2004.

J. K. G. Watson. Vibration-rotation hamiltonians of linear-molecules. *Mol. Phys.*, 79:943–951, 1993.

J. K. G. Watson. Robust weighting in least-square fits. *J. Mol. Spectrosc.*, 219:326–328, 2003.

H. Wei and T. Carrington. Explicit expressions for triatomic Eckart frames in Jacobi, Radau, and bond coordinates. *J. Chem. Phys.*, 107:2813–2818, 1997.

H. Wei and T. Carrington Jr. An exact Eckart-embedded kinetic energy operator in Radau coordinates for triatomic molecules. *Chem. Phys. Lett.*, 287:289–300, 1998.

E. B. Wilson, J. C. Decius, and P. C. Cross. *Molecular Vibrations: The Theory of Infrared and Raman Vibrational Spectra.* McGraw-Hill, New York 1955.

G. Worth. Quantics: A general purpose package for Quantum molecular dynamics simulations. *Comput. Phys. Commun.*, 248:107040, 2020.

D. G. Xu, G. H. Li, D. Q. Xie, and H. Guo. Full-dimensional quantum calculations of vibrational energy levels of acetylene (HCCH) up to 13,000 cm^{-1}. *Chem. Phys. Lett.*, 365:480–486, 2002.

A. Yachmenev and S. N. Yurchenko. Automatic differentiation method for numerical construction of the rotational-vibrational Hamiltonian as a power series in the curvilinear internal coordinates using the Eckart frame. *J. Chem. Phys.*, 143:014105, 2015.

W. Yang and A. C. Peet. The collocation method for bound solutions of the Schrödinger equation. *Chem. Phys. Lett.*, 153(1):98–104, 1988.

H.-G. Yu. An exact variational method to calculate vibrational energies of five atom molecules beyond the normal mode approach. *J. Chem. Phys.*, 117(5):2030–2037, 2002.

H.-G. Yu. An exact variational method to calculate rovibrational spectra of polyatomic molecules with large amplitude motion. *J. Chem. Phys.*, 145: 084109, 2016.

H. G. Yu and J. T. Muckerman. A general variational algorithm to calculate vibrational energy levels of tetraatomic molecules. *J. Mol. Spectrosc.*, 214: 11–20, 2002.

S. N. Yurchenko and T. M. Mellor. Treating linear molecules in calculations of rotation-vibration spectra. *J. Chem. Phys.*, 153(15):154106, 2020.

S. N. Yurchenko and J. Tennyson. ExoMol line lists - IV. The rotation-vibration spectrum of methane up to 1500 K. *Mon. Not. R. Astron. Soc.*, 440:1649–1661, 2014.

S. N. Yurchenko, M. Carvajal, P. Jensen, F. Herregodts, and T. R. Huet. Potential parameters of PH_3 obtained by simultaneous fitting of ab initio data and experimental vibrational band origins. *Chem. Phys.*, 290:59–67, 2003.

S. N. Yurchenko, M. Carvajalz, P. Jensen, H. Lin, J. J. Zheng, and W. Thiel. Rotation-vibration motion of pyramidal XY_3 molecules described in the Eckart frame: Theory and application to NH_3. *Mol. Phys.*, 103:359–378, 2005a.

S. N. Yurchenko, W. Thiel, M. Carvajal, H. Lin, and P. Jensen. Rotation-vibration motion of pyramidal XY_3 molecules described in the Eckart frame: The calculation of intensities with application to NH_3. *Adv. Quant. Chem.*, 48:209–238, 2005b.

S. N. Yurchenko, W. Thiel, and P. Jensen. Theoretical ROVibrational Energies (TROVE): A robust numerical approach to the calculation of rovibrational energies for polyatomic molecules. *J. Mol. Spectrosc.*, 245:126–140, 2007.

S. N. Yurchenko, R. J. Barber, and J. Tennyson. A variationally computed line list for hot NH_3. *Mon. Not. R. Astron. Soc.*, 413:1828–1834, 2011.

S. N. Yurchenko, L. Lodi, J. Tennyson, and A. V. Stolyarov. Duo: A general program for calculating spectra of diatomic molecules. *Comput. Phys. Commun.*, 202:262 – 275, 2016.

S. N. Yurchenko, A. Yachmenev, and R. I. Ovsyannikov. Symmetry adapted ro-vibrational basis functions for variational nuclear motion: TROVE approach. *J. Chem. Theory Comput.*, 13(9):4368–4381, 2017.

S. N. Yurchenko, T. M. Mellor, R. S. Freedman, and J. Tennyson. ExoMol line lists – XXXIX. Ro-vibrational molecular line list for CO_2. *Mon. Not. R. Astron. Soc.*, 496(4):5282–5291, 2020.

R. N. Zare. *Angular Momentum: Understanding Spatial Aspects in Chemistry and Physics.* 1st edition, Wiley, New York, 1988.

D. H. Zhang, Q. Wu, J. Z. H. Zhang, M. von Dirke, and Z. Bačić. Exact full-dimensional bound state calculations for $(HF)_2$, $(DF)_2$, and HFDF. *J. Chem. Phys.*, 102:2315–2325, 1995.

B. I. Zhilinskii, V. I. Perevalov, and V. G. Tiuterev. *The Method of Irreducible Tensor Operators in the Theory of Molecular Spectra.* Nauka, Novosibirsk, 1987.

Index